SpringerBriefs in Environmental Science

For further volumes:
http://www.springer.com/series/8868

Winfried Schröder · Gunther Schmidt

Modelling Potential Malaria Spread in Germany by Use of Climate Change Projections

A Risk Assessment Approach Coupling Epidemiologic and Geostatistical Measures

 Springer

Winfried Schröder
Gunther Schmidt
Lehrstuhl für Landschaftsökologie
Universität Vechta
Vechta
Germany

ISSN 2191-5547   ISSN 2191-5555 (electronic)
ISBN 978-3-319-03822-3   ISBN 978-3-319-03823-0 (eBook)
DOI 10.1007/978-3-319-03823-0
Springer Cham Heidelberg New York Dordrecht London

Library of Congress Control Number: 2013955558

Printed on acid-free paper

Springer is part of Springer Science+Business Media (www.springer.com)

Contents

Abstract

According to estimated future temperature rise of the Intergovernmental Panel on Climate Change (IPCC), it has to be expected that the spatial and temporal extent for malaria transmission in Germany will increase as well. Climate warming can affect the distribution and the intensity of parasitic diseases that are carried by insects and animals (vector-borne diseases). This is because the parasites that cause the disease usually flourish in increased temperatures where they benefit from accelerated rates of reproduction and development. Malaria is usually thought to be restricted to the tropics and developing countries, but climate change could bring it back to Europe, especially into countries where it was present until the middle of the last century, such as, for example, Germany: tertian malaria or vivax malaria, a rather severe form of malaria, was prevalent in north-western parts of Germany until the 1950s before it was eradicated. The vector itself (the mosquito) is still present and infected people from malarial regions could introduce a new onset of malaria.

The study at hand comprises two case studies: in *case study 1* areas at risk of a malaria outbreak in the German federal state Lower Saxony, a former endemic malaria region until the 1950s, were mapped. Subsequently, the study was broadened to the whole territory of Germany (*case study 2*). *Case study 1* was based on measured (1947–1960, 1961–1990, 1985–2004) and predicted (2020, 2060, 2100, each best case and worst case scenario) air temperatures. In *case study 2*, the potential spread of tertian malaria by anopheles mosquitoes was modelled for the periods 1991–2020, 2021–2050, and 2051–2080 using the IPCC A1 and B1 emission scenarios. Both studies rely on data from the relevant literature defining the reproduction and development of the mosquito (*Anopheles atroparvus*) and the malarial parasite as considered in the basic reproduction rate (R_0) formula.

The results from *case study 1* indicated that the gate of potential tertian malaria transmissions with regard to R_0 could be expected to increase from 2 months in the past to 6 months in the future in Lower Saxony. Past and recent findings of *Anopheles atroparvus* coincide with those regions where the potential malaria transmission gate was projected lasting for 4 months in 2060 (best case scenario) and for 6 months in 2100 (worst case scenario) and, in addition, where tertian malaria occurred up to the 1950s. From *case study 2*, it could be derived that, amongst others, between 1991 and 2007 about 70 % of Germany, mainly in coastal regions and foothills, had a transmission gate of 3 months. Taking the temperature estimates for 2051–2080 into account, up to 96.5 % of Germany could experience a 4–5 month transmission window.

Evaluating these results, it has to be empasized that the models offer, at the best, an approximation to future reality since not all potentially relevant factors could have been included into the computations. The predictions focused on changes in air temperature and did not consider other influences, such as the number and geographical distribution of water bodies which are breeding habitats for mosquitoes, population density, or other socio-economic factors. There may also be interactions between the mosquitoes, parasites and features of the ecosystem, such as rainfall and soil temperature, as well as community hygiene. Modeling studies, such as this investigation, can help to identify areas at risk with the aim of initiating preventative measures, especially if they include the factors mentioned above. In particular, the presented approach produced new outcome by coupling epidemiological methods, i.e., application of the basic reproduction rate (R_0) for *Anopheles atroparvus*, geostatistical measures provided by Geographical Information Systems (GIS) and modelled temperature projections based on IPCC scenarios. The presented approach could also be applied to other vector-borne diseases that are dangerous for livestock, such as bluetongue disease or the Akabane disease transmitted by Culicoides.

Keywords *Anopheles atroparvus* · Basic reproduction rate · Epidemiologic climate change impacts · Geographical Information System · Geostatistics · Malaria tertiana · Modeling and mapping · *Plasmodium vivax* · Vector-borne diseases

Chapter 1
Background and Goals

1.1 Background

Malaria is a life-threatening disease caused by parasites that are transmitted to people through the bites of infected mosquitoes (WHO 2013). According to the latest estimates, there were about 219 million cases of malaria in 2010 with an uncertainty range of 154–289 million and 660,000 deaths with an uncertainty range of 490,000–836,000. Malaria can be prevented and cured: Malaria mortality rates decreased by more than 25 % globally since 2000 and by 33 % in the WHO African Region. Non-immune travellers from regions currently free from endemic malaria, such as Germany, are very vulnerable to the disease when they get infected (Behrens et al. 2010).

Malaria is caused by Plasmodium parasites. The parasites are spread to people through the bites of infected Anopheles mosquitoes, called malarial vectors. Four parasite species cause malaria in humans: *Plasmodium falciparum* (Malaria tropica)*, Plasmodium malariae* (Malaria quartana), *Plasmodium vivax* (Malaria vivax/Malaria tertiana) and *Plasmodium ovale* (Malaria tertiana). *P. falciparum* and *P. vivax* are the most common malaria parasites, *P. falciparum* is the most severe.

Malaria is transmitted through the bites of Anopheles mosquitoes. The number of transmissions depends on factors related to the parasite, the vector, the human host and the environment. Transmission is more intense in places where the mosquito lifespan is longer, so that the parasite has time to complete its development inside the mosquito (Sainz-Elipe et al. 2010). Climatic conditions such as precipitation patterns, temperature and humidity may affect the number and survival of mosquitoes and, thus, transmission. Malaria epidemics can occur when climate and other conditions favour transmission in areas where people have little or no immunity to malaria. They can also occur when people with low immunity move into areas with intense malaria transmission, for instance, European tourists. Thus, specific population risk groups include, amongst others, international travellers from non-endemic areas because they lack immunity. Immigrants from endemic areas and their children living in non-endemic areas and returning to their

W. Schröder and G. Schmidt, *Modelling Potential Malaria Spread in Germany by Use of Climate Change Projections*, SpringerBriefs in Environmental Science, DOI: 10.1007/978-3-319-03823-0_1, © The Author(s) 2014

home countries to visit friends and relatives are similarly at risk because of waning or absent immunity (Yukich et al. 2013). Two types of mobility patterns can be identified that promote the spread of infections mainly: migration, i.e. when the people move from one region to another with no returns or when people return to their country of origin after visiting other regions (Mandal et al. 2011). As of 2010, about 100 countries worldwide are endemic for malaria (WHO 2012). Every year, 125 million international travellers visit these countries, and more than 30,000 contract the disease (Kajfasz 2009). In Germany, about 900 malaria cases were registered between 1980 and 2006, and 0.3–0.9 % of the people died. Most of the German tourists suffering from malaria travelled through malaria endemic regions of Africa (87 %).

Considering the increased mobility of humans between Europe and malaria endemic countries (Askling et al. 2012) and changes in environment and climate, especially the knowledge of the transmission intensity and climatic factors, it has to be clarified whether climate change (warming up) could increase the malaria transmission in Europe (Zoller et al. 2009). Over the last 15 years, autochthonous cases and epidemic outbreaks of vector-borne diseases have occurred in Europe (e.g. bluetongue, chikungunya, dengue, Schmallenberg disease, Usutu virus infection), including local transmission of malaria in France, Germany, Greece, Italy and Spain. Driving factors contributing to the emergence or re-emergence of vector-borne diseases are the increased mobility of humans, livestock and pathogens as well as agronomical, environmental and climatic changes. This emphasizes the need to evaluate or re-evaluate the capability of European Anopheline mosquitoes to allow local transmission of malaria (Edlund et al. 2012).

Autochthonous transmission of malaria depends on the presence of the Anopheles vector, infectious persons, individuals susceptible for malaria, and favourable climatic conditions. In Germany, these conditions and, thus, a potential for autochthonous malaria transmission are assumed to be present: The risk of autochthonous malaria is suspected to increase due to global warming and triggered by the increasing number of malaria cases by infested travellers (Dalitz 2005). Epidemiologists were one of the foremost to realise the important role of mathematical modelling as a tool for better understanding the disease transmission dynamics within and between hosts and parasites and their respective environments (Mandal et al. 2012). The same holds true for the application of Geographical Information Systems (GIS) and spatial analysis. Despite serious objections preventing from overestimating the relevance of predicting the potential influence of global warming to the transmission and epidemiology of malaria (Reiter 2008), today, risk modelling is one of the precautionary means in the framework of public health and management of malaria, especially in view of climate change (Dalitz 2005). Warming of the climate system is indisputable, as is evidenced by, amongst others, measurements of air temperatures: Eleven of the years 1995–2006 rank among the twelve warmest years in the instrumental record of global surface temperature since 1850. The 100-year linear trend (1906–2005) amounted for in average 0.74 °C. The temperature increase is a global phenomenon and more intense at higher northern latitudes (IPCC 2007). Two distinct periods of temperature increase could be identified: The first took place from 1910 to 1945 and the second could be determined since 1976. Within both periods, the 1990s have been the warmest decade since the beginning of meteorological observation, and in Germany, 1998 was the warmest year ever observed between 1861 and 2000 (IPCC 2001). The globally averaged surface temperature measurements indicate a temperature increase of about 0.2 °C per decade, and comparing the periods 2090–2099 and 1980–1999 the climate projections yield estimates of a potential temperature increase of about 1.1–6.4 °C according to the respective region and emission scenario (IPCC 2007). For whole Germany, the average increase in air temperatures between 1991 and 2002 was about 0.8 °C with regional peaks of up to 3.5 °C (Schröder et al. 2010). In strong correlation with this warming, a shift in the onset of phenological phases of plants and animals, especially in spring, can be ascertained as well as a change in their geographical distribution to higher latitudes and altitudes. And there is medium confidence that other effects of regional climate change on natural and human environments are emerging: alterations in disturbance regimes of forests due to fires and pests as well as changes in infectious disease vectors in some areas (IPCC 2007).

It is argued that malaria research and management need a holistic view of those factors primarily influencing the transmission and epidemiology of the disease: the ecology and behaviour of both humans and vectors as well as the immunity of the human population (Reiter 2008). These factors should be assumed to interact within a network of highly variable parameters. Thus, the dynamics of transmission could only be assessed by taking this complexity into account. This includes that the key climate variables such as air temperatures, precipitation and humidity cannot be viewed independently: For example, the effects of temperature are modified by humidity: the daily range of each element may be more significant than the daily mean. Or brief periods of atypical heat or cold can be more significant than long-term averages. The significance of these variables, and their putative role in future climates, should be assessed in the perspective of this complexity (Reiter 2008). But would it be enough to cover manifold biological, chemical and physical variables in malaria research and management? An analysis of indigenous malaria in Finland evaluated the significance of factors assumed to affect malaria trends. It could be shown that the indigenous malaria in Finland faded out evenly in the whole country during 200 years with limited or no countermeasures or medication. It could be concluded that malaria in Finland was basically a social disease and that malaria trends were strongly linked to changes in human behaviour. Decreasing household size caused fewer interactions between families and accordingly decreasing re-colonisation possibilities for Plasmodium. The decrease of the household size was the precondition for a permanent eradication of malaria (Hulden and Hulden 2009).

1.2 Goals

Hoshen and Morse (2005) classify malaria transmission models into statistical, rule-based and dynamic models. Accordingly, the statistical models compare malaria transmission variables with local conditions (Killeen et al. 2000). The rule-based models determine the regions in which malaria transmission is possible (Snow et al. 1998), and the dynamic models relate malaria transmissions to constant climate conditions (Bailey 1982). Models coupling malaria transmission and changing climate serving for pre-emptive decision-making based on climate change scenarios seemed to be missing. In view of the recommendations given by Gemperli et al. (2006), Gimnig et al. (2005), Hendrickx et al. (2004), Kleinschmidt et al. (2000), Martens and Thomas (2005) and WHO (2004), this investigation should help to partially fill this gap by combining a mathematical model of the reproduction rate of *P. vivax* transmitted by *Anopheles atroparvus* with a multivariate ecological land classification, geostatistical surface estimation and a Geographic Information System (GIS) (Schröder and Schmidt 2008). Given this, the background information reported above and serious considerations regarding potential malaria outbreaks in Germany with increasing air temperatures (Dalitz 2005), we conducted two case studies within one methodical framework (Schröder 2006): The first one was designed as a pilot study referring to Lower Saxony, one of the 16 federal states of Germany and an endemic malaria region until the 1950s. Based on the results yielded from this pilot study (Schmidt et al. 2008) we extended the investigation to the whole territory of Germany (Holy et al. 2009a, b, 2011; Schröder et al. 2010). The results from both case studies were published in peer reviewed journals and are compiled in this issue of *SpringerBriefs* in Environmental Sciences.

The area under investigation in *case study* 1, Lower Saxony (Northwest Germany), is a former malaria region with highest incidences along the coastal zones. Malaria had finally become eradicated in the early 1950s. Subsequently, further scientific investigations in that field declined. Nevertheless, Anopheles mosquitoes which transmit malaria have been observed in Lower Saxony until the beginning of the twenty-first century. Thus, the question arose whether a new autochthon transmission could take place if the pathogen is introduced again and could develop in Anopheles mosquitoes. Answering this question was the first aim of *case study* 1. The second one was to examine the spatial and temporal structure of potential transmissions in respect to the predicted increase in the air temperatures according to the IPCC scenarios. To answer these questions, current information on Anophelinae and their distribution and habitat preferences were collected by literature research. Complementarily, temperature measurements and Anopheles findings were compiled from the Deutscher Wetterdienst (German Weather Survey) and the Niedersächsisches Landesamt für Ökologie (Lower Saxony Authority for Environmental Affairs), respectively. Thus, the spatial distribution of potential temperature-driven malaria transmissions could be investigated using both, the basic reproduction rate (R_0) for calculating the risk of an outbreak of tertian malaria based on measured (1947–1960, 1961–1990, 1985–2004) and predicted (2020, 2060, 2100, each best-case and worst-case scenario) monthly mean air temperature data, and geostatistics for mapping respective areas at risk. The resulting risk maps were then intersected with a map on ecologically defined land units of Lower Saxony within a GIS environment. From the results of *case study* 1 (Lower Saxony) could be concluded that the approach applied enabled risk calculation and mapping for different ecologically defined land classes at the regional level.

The aim of *case study* 2 was to adopt this approach by applying it to an area of larger spatial extent, i.e. for the whole territory of Germany. To this end, the spatial distribution of potential temperature-driven malaria transmissions was calculated using the basic reproduction rate (R_0) applied for measured (1961–1990, 1991–2007) and predicted (1991–2020, 2021–2050, 2051–2080 each IPCC A1B and B1 scenario) monthly mean air temperature data. The mapping of areas at risk of a malaria outbreak was spatially differentiated by use of an ecological land classification for the whole territory of Germany.

References

Askling HH, Bruneel F, Burchard G, Castelli F, Chiodini PL, Grobusch MP, Lopez-Vélez R, Paul M, Petersen E, Popescu C, Ramharter M, Schlagenhauf P (2012) Management of imported malaria in Europe. Malar J 11:328

Bailey NTJ (1982) The Biomathematics of Malaria. Griffin, London

Behrens RH, Carroll B, Hellgren U, Visser LG, Siikamäki H, Vestergaard LS, Calleri G, Jänisch T, Myrvang B, Gascon J, Hatz C (2010) The incidence of malaria in travellers to South-East Asia: is local malaria transmission a useful risk indicator? Malar J 9:266

Dalitz MK (2005) Autochthone Malaria im mitteldeutschen Raum. Dissertation, Martin-Luther-Universität, Halle-Wittenberg

Edlund S, Davis M, Douglas JV, Kershenbaum A, Waraporn N, Lessler J, Kaufman JH (2012) A global model of malaria climate sensitivity: comparing malaria response to historic climate data based on simulation and officially reported malaria incidence. Malar J 11:331

Gemperli A, Vounatsou P, Sogoba N, Smith T (2006) Malaria mapping using transmission models. Application to survey data from Mali. Am J Epidemiol 163:289–297

Gimnig JE, Hightower AW, Hawley WA (2005) Application of geographic information systems to the study of the ecology of mosquitoes and mosquito-borne diseases. In: Takken W, Martens P, Bogers RJ (eds) Environmental change and malaria risk. Global and local implications. Springer, Dordrecht

Hendrickx G, Biesemans J, de Deken R (2004) The use of GIS in veterinary parasitology. In: Durr PA, Gatrell AC (eds) GIS and spatial analysis in veterinary science. CABI Publishing, Wallingford

Holy M, Pesch R, Schmidt G, Schröder W (2009a) Aufbau eines Fachinformationssystems "Klimafolgen und Anpassung" (FISKA). FuE-Vorhaben im Auftrag des Umweltbundesamtes (FKZ 20641100), Abschlussbericht (14 Sep 2009). Umweltbundesamt, Dessau

Holy M, Schmidt G, Schröder W (2009b) GIS-basierte Risikomodellierung zur Auswirkung des Klimawandels auf die potenzielle Malariaausbreitung in Deutschland. In: Strobl J, Blaschke T, Griesebner G (eds) Angewandte Geoinformatik 2009. Wichmann, Heidelberg

Holy M, Schmidt G, Schröder G (2011) Potential malaria outbreak in Germany due to climate warming: risk modelling based on temperature measurements and regional climate models. Environ Sci Pollut Res 18:428–435

Hoshen MB, Morse AP (2005) A model structure for estimating malaria risk In: Takken W, Martens P, Bogers RJ (eds) Environmental change and malaria risk: global and local implications. Springer, Dordrecht

Hulden L, Hulden L (2009) The decline of malaria in Finland—the impact of the vector and social variables. Malar J 8:94

IPCC (Intergovernmental Panel of Climate Change) (2001) Climate change. The scientific basis. Cambridge University Press, Cambridge

IPCC (Intergovernmental Panel of Climate Change) (2007) Climate change 2007. Synthesis report, Geneva

Kajfasz P (2009) Malaria prevention. Int Marit Health 60(1–2):67–70

Killeen GF, McKenzie FE, Foy BD, Schieffelin C, Billingsley PF, Beier JC (2000) A simplified model for predicting malaria entomologic inoculation rates based on entomologic and parasitologic parameters relevant to control. Am J Trop Med Hyg 62:535–544

Kleinschmidt I, Bagayoko M, Clarke GPY, Craig M, Le Sueur D (2000) A spatial statistical approach to malaria mapping. Int J Epidemiol 29:355–361

Mandal S, Sarkar RR, Sinha S (2011) Mathematical models of malaria—a review. Malar J 10:202

Martens P, Thomas C (2005) Climate change and malaria risk: complexity and scaling. In: Takken W, Martens P, Bogers RJ (eds) Environmental change and malaria risk. Global and local implications. Springer, Dordrecht

Reiter P (2008) Global warming and malaria: knowing the horse before hitching the cart. Malar J 7(1):S3

Sainz-Elipe S, Latorre JM, Escosa R, Masià M, Fuentes MV, Mas-Coma S, Bargues MD (2010) Malaria resurgence risk in southern Europe: climate assessment in an historically endemic area of rice fields at the Mediterranean shore of Spain. Malar J 9:221

Schmidt G, Holy M, Schröder W (2008) Vector-associated diseases in the context of climate change: Analysis and evaluation of the differences in the potential spread of tertian malaria in the ecoregions of Lower Saxony. Ital J Public Health 5(4):245–252

Schröder W (2006) GIS, geostatistics, metadata banking, and tree-based models for data analysis and mapping in environmental monitoring and epidemiology. Int J Med Microbiol 296(40):23–36

Schröder W, Schmidt G (2008) Mapping the potential temperature-dependent tertian malaria transmission within the ecoregions of Lower Saxony (Germany). Int J Med Microbiol 298(S1):38–49

Schröder W, Holy M, Pesch R, Schmidt G (2010) Klimawandel und zukünftig mögliche temperaturgesteuerte Malariatransmission in Deutschland. Umweltwiss Schadst Forsch Z Umweltchem Ökotox 22:177–187

Snow RW, Gouws E, Omumbo J, Rapuoda B, Craig MH, Tanser FC, le Suer D, Ouma J (1998) Models to predict the intensity of Plasmodium falciparum transmission: applications to the burden of disease in Kenya. Trans Roy Soc Trop Med H 92:601–606

WHO (World Health Organistion) (2004) Using climate to predict infectious disease outbreaks. A review, Geneva

WHO (World Health Organisation) (2012) World Malaria report 2012. World Health Organization, Geneva, Switzerland

WHO (World Health Organisation) (2013) WHO Malaria fact sheet N°94 January 2013. http://www.who.int/mediacentre/factsheets/fs094/en/index.html. Accessed on 18 June 2013

Yukich JO, Taylor C, Eisele TP, Reithinger R, Nauhassenay H, Berhane Y, Keating J (2013) Travel history and malaria infection risk in a low-transmission setting in Ethiopia: a case control study. Malar J 12:33

Zoller T, Naucke TJ, May J, Hoffmeister B, Flick H, Williams CJ, Frank C, Bergmann F, Suttorp N, Mockenhaupt FP (2009) Malaria transmission in nonendemic areas: case report, review of the literature and implications for public health management. Malar J 8:71

Chapter 2
Case Study 1: Modelling Potential Transmission Gates of Malaria Tertiana in Lower Saxony

Abstract It is widely assumed that climate change is likely to affect the geographic distribution and intensity of the transmission of vector-borne diseases such as malaria. These diseases are expected to occur, compared with the past and presence, at higher latitudes and altitudes. A slight rise in ambient temperature and precipitation is expected to extend the duration of the season in which mosquito vectors are transmitting the malaria pathogens. The parasites as well usually benefit from increased temperatures, as both their reproduction and development would accelerate. These inter-relationships and respective public discussions gave reason to examine potential effects on the seasonal transmission gate due to the predicted climate changes in Lower Saxony (North-Western Germany). The federal state Lower Saxony was a former endemic malaria region with highest incidences of *Anopheles atroparvus* and tertian malaria along the coastal zones until malaria had finally been eradicated in the early 1950s. However up to now, the malaria vector Anopheles is still present. Accordingly, a pilot study should settle whether a new autochthonous transmission could take place if the malaria pathogen is introduced again in Lower Saxony. Thus, the spatial and temporal structure of temperature-driven malaria transmissions was investigated using the basic reproduction rate (R_0) to model the risk for an outbreak of tertian malaria due to measured (1947–1960, 1961–1990, 1985–2004) and predicted (2020, 2060, 2100, each best case and worst case scenario) air temperatures and to geostatistically map the respective risk areas. The risk maps revealed that the gate of potential tertian malaria transmissions in terms of R_0 could be expected to increase from two months in the past to six months in the future in Lower Saxony. Past and recent findings of *A. atroparvus* coincide with those regions where the potential malaria transmission gate accounts for four months in 2060 (best case scenario) and for six months in 2100 (worst case scenario) and, in addition, where tertian malaria occurred up to the 1950s. The estimated maps on malaria risk were intersected with a map on ecological land units, enabling an ecological regionalisation.

Keywords *Anopheles atroparvus* · Basic reproduction rate · Epidemiologic climate change impact · Malaria tertiana · *Plasmodium vivax*

W. Schröder and G. Schmidt, *Modelling Potential Malaria Spread in Germany by Use of Climate Change Projections*, SpringerBriefs in Environmental Science, DOI: 10.1007/978-3-319-03823-0_2, © The Author(s) 2014

2.1 Background and Goals

Malaria is one of the major causes of global mortality and morbidity even though the aetiology of the disease has been known for many years (Ross 1911). Public health related management of malaria is assumed to be supported by enhanced application of existing knowledge. Modelling approaches could help combining various sources of data regarding different aspects of the disease dynamics and link these aspects with external causes or covariates serving as driving forces or as surrogates for them.

The latest incidences of autochthonous malaria in Germany were reported in the early 1950s (Weyer 1956). However, in the face of current climate change, research on vector-borne diseases and investigations on the population density and distribution of Anopheles were conducted in some European countries (Eritja et al. 2000; Kubica-Biernat 1999; Romi et al. 1997; Schaffner 1998). For Germany only poor data sets were available (Maier et al. 2003). This gap together with the rise of air temperatures and the correlated ecological changes as indicated by, e.g., plant phenology throughout Germany (Schröder et al. 2005) gave reason to investigate whether the former malaria region Lower Saxony could be at risk of a new outbreak and to what extent higher temperatures could prolong the potential transmission gate of the tertian malaria pathogen.

2.2 Methods

2.2.1 Literature Research on Malaria Transmission

The estimation of the geographical patterns for the incidence of malaria vectors presumes that their correlation with habitat characteristics and their life stages can be quantitatively described. To this end literature research was conducted. The following information was initially researched and used for the GIS-based risk modelling of a climate warming induced malaria tertiana outbreak in Lower Saxony. Later on, the same methodology was applied to the German-wide investigation, too (Sect. 2.3).

Approximately 3,500 species of mosquitoes grouped into 41 genera are known and about 30–40 of about 430 Anopheles species transmit malaria. From these, six species could be identified in Germany. Females of the genus *Anopheles* (Diptera, Culicidae) transmit tertian malaria (Ramsdale and Snow 2000). In Lower Saxony, *A. atroparvus* was dominantly associated with malaria cases (Hackett and Missiroli 1935; Martini 1920a, b; Weyer 1940). In this study, all historical Anopheles findings and malaria cases documented in literature were localised and mapped in a GIS. They were supplemented by *Anopheles* findings since 1985 documented in the digital so-called BOG-archive (Biologische Oberflachengewässer) held by the Lower Saxony Authority for Environmental Affairs (Schröder et al. 2007) or published by Wilke et al. (2006) (Sect. 2.2.2).

Apart from several factors that affect the aquatic juvenile stages of *A. atro-parvus*, one of the main driving forces for development is air temperature. Jetten and Takken (1994) compiled data on the required time for development and related temperature values: the lower development threshold, the optimum temperature, and the upper limit above which no further progress is possible. This information was essential for the identification of areas at risk. As mainly climate conditions influence the development of the Anophelinae, the number of generations growing in one season varies spatially. Martini (1946) estimated 5–7 generations as an average value for Central Italy, 2–4 generations for Southern Germany and 2–3 for Northern Germany (Heinz 1950). In the summer of 1947 showing unusually high temperatures in north-western Germany, the growing of up to 5 generations could have been possible (Heinz 1950).

Malaria is caused by infection with an intracellular protozoan parasite of the genus *Plasmodium*. Human beings are commonly infected by *P. falciparum*, *P. vivax*, *P. ovale*, or *P. malariae*. Their development in the vector mosquito is predominantly influenced by temperature. *P. vivax* causes tertian malaria, which was dominant in north-western Germany (Mühlens 1930). A recently infected mosquito needs a period of at least 105 days at 14.5 °C, the lower thermal threshold for the development of *Plasmodium* in mosquitoes, to get infectious. This period may be interrupted by some colder days (Jetten and Takken 1994). Concerning both case studies, the identification of the high-risk areas was performed by data referring to *P. vivax* given by Jetten and Takken (1994).

2.2.2 Data Compilation in a Geographic Information System

Data processing and risk assessment were based on data compiled from several sources and integrated into a Geographic Information System (GIS): From literature, information on historical *Anopheles* findings were collected (Sect. 2.2.1). These included information on the taxonomy of the respective species, the name of the locality, the Gauß-Krüger (transverse Mercator projection) coordinates and the literature source, respectively. Referring to the recent (since 1985) findings of the Lower Saxony Authority for Environmental Affairs documented in the "BOG-Archiv", the same metadata were collected and supplemented with additional details about the name of the water body and the date the *Anopheles* larvae were found.

Air temperature data from 54 observation sites in Lower Saxony were provided by the German Weather Survey in terms of diurnal extreme values that were aggregated to monthly means of air temperature. Taking into account that in temperate climatic zones such as Northern Germany a successful transmission of the malaria pathogen is only possible in the warmest months of the year, this study focusses on the period from June to August for past and present analyses. September and May were not included for processing the past periods because the low average temperatures did not allow pathogen transmission. According to the predicted global warming, the calculations for future trends were also performed

for these two months and, additionally, October. For data processing, several data
sets were generated which list the monthly means of temperature of June, July or
August, respectively, of the following periods: 1947 1960, 1961–1990 and
1985 2004. The beginning of the latter period was set to 1985 instead of 1991
because the BOG-Archiv starts with respective findings of Anophelinae larvae in
1985. By this, a complete temporal link between the data on both, the air temperatures and the findings of potential vectors, was assured.

The site-specific meterological data were geostatistically transformed to surface
data (Sect. 2.2.3) and spatially differentiated in terms of ecologically defined land
units (Sect. 2.2.4). The geostatistical surface estimations enabled the spatial link of
the air temperature data with the Anopheles findings and associated malaria
incidences until the early 1950s and with the Anopheles findings documented in
the literature and the BOG-archive since 1985 (Sect. 2.2.1).

2.2.3 Geostatistical Estimation of Temperature Maps

Since the measurement data from the 54 meteorological observation sites
(1947–1960, 1961–1990, 1985–2004) did not match spatially with both the
Anopheles findings and the malaria cases, we used the geostatistical tool FUZZ-
EKS (Bartels 1997; Piotrowski et al. 1996) to derive maps on air temperature
conditions in Lower Saxony for the different periods of investigation. Geostatistics
enable to analyse and model the spatial autocorrelation of biological, chemical and
physical measurements, and, based on this variographic modelling, to calculate
surface estimations by spatially weighted kriging interpolation (Webster and
Oliver 2001). The quality of the estimation was proved by cross-validation: Each
measurement was sequentially extracted from the sample and was then estimated
by kriging according to the variographic model. The differences between measured
and geostatistically estimated values were then statistically investigated yielding
good results. This approach enables to correlate spatially incongruent data, such as
air temperature measurements, Anopheles findings, malaria cases, and data on
landscape characteristics from a map on ecological landclasses (Sect. 2.2.4).

In addition, projected monthly air temperature means (IPCC 2001) for the years
2020, 2060 and 2100 were calculated in the GIS environment. For each of these
years, two different climate predictions were used: the "best case (b.c.)" represents
the lowest predicted temperature values, the "worst case (w.c.)", by contrast,
reflects the highest possible temperature values calculated in the 35 climate scenarios published by IPCC (2001): 2020 b.c. +0.3 °C, w.c. +0.9 °C; 2060
b.c. +0.9 °C, w.c. +3.3 °C; 2100 b.c. +1.4 °C, w.c. +5.8 °C). For each period and
scenario, the predicted values were added to the temperature grid of the period
1985–2004 to get according maps for 2020, 2060, and 2100, one indicating the
best and one another indicating the worst case scenario. These projected air
temperature data as well as the measured values covering the years 1947–2004
were used as input for the calculation of the basic reproduction rate R_0 (Sect.
2.2.5).

2.2.4 Ecological Land Classification

Ecological land classifications were computed based on data which represent several interacting factors which may be of importance for natural processes such as the transmission of malaria. Hence, consideration of maps depicting the spatial structure of complex interacting ecological factors for the modelling of potential malaria transmission covers at least partly the requirements postulated by Reiter (2008). The ecological land classification used in both case studies was calculated from the data in Table 2.1 applying Classification and Regression Trees (CART; Breiman et al. 1984). The respective computations yielded maps (Fig. 2.1) illustrating the spatial patterns of georeferenced ecologically defined land classes which synonymously are called natural land classes, ecoregions or landscapes (Schmidt 2002; Schröder 2006; Schröder et al. 2006). All ecoregions are itemised with regard to the values or the statistical distribution of values of 48 ecological characteristics which were used to generate the land classes by use of CART (Schröder and Schmidt 2001).

Table 2.1 Data used for ecological land classification (according to Schröder and Schmidt 2008)

Map title	Observation period/State	GIS layer	Source
Potential natural vegetation	1998	1	BfN
Soil texture	2000	1	BGR
Altitude above sea level	1996	1	UNEP
Mean of monthly global radiation March–November	1981–1999	9	DWD
Mean of monthly evaporation January–December	1961–1990	12	DWD
Mean of monthly precipitation January–December	1961–1990	12	DWD
Mean of monthly air temperature January–December	1961–1990	12	DWD

BfN Federal Agency for Nature Conservation, *BGR* Federal Institute for Geosciences and Natural Resources, *GIS* Geographical Information System, *UNEP* United Nations Environmental Programme, *DWD* German National Meteorological Service

2.2.5 Calculation of the Basic Reproduction Rate (R_0)

The geostatistically calculated surface air temperature maps were used for GIS-based risk modelling of a climate warming induced outbreak of malaria tertiana in Lower Saxony. Traditional approaches measuring whether malaria transmission has the potential for epidemics have focused on the bio-mathematical relationships between the parasite and its primary host, the mosquito (Smith and McKenzie 2004). These relationships have been quantified, e.g., as the stability index, basic reproduction rate and vectorial capacity (Snow et al. 1999). In the investigation at hand, the basic reproduction rate (R_0) was used to analyse whether areas at risk for

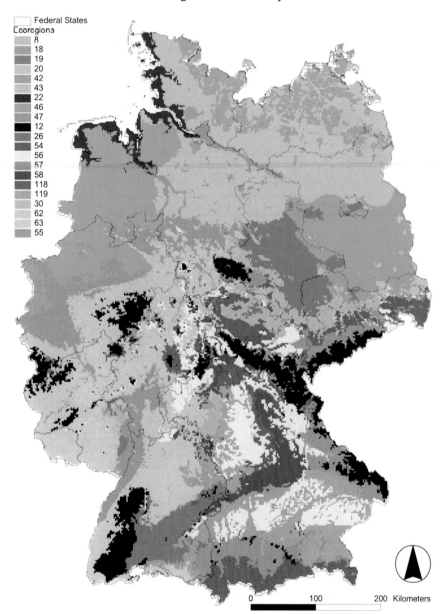

Fig. 2.1 Eco-regions of Germany calculated by CART (Classification and regression trees) from the data in Table 2.1 (according to Schröder and Schmidt 2001)

an outbreak of malaria tertiana can be found in Lower Saxony due to its temperature conditions. Considering the malaria protozoon, R_0 could be more precisely defined as the average number of secondary infections occurring when a single infected individual is introduced into a potential host population in which each member is susceptible. If $R_0 \geq 1$, the disease will proliferate indefinitely, if $R_0 < 1$, the disease will die out (Martens et al. 1999: 92). The original formula was altered and specified over the years by several authors, but its function remains the same. The R_0-formula used for this analysis is shown and explained below:

$$\frac{m \cdot a^2 \cdot b \cdot p^n}{-\ln(p) \cdot r}$$

m	relative density of female Anophelines
a	frequency of feeding on a person, expressed as a daily rate
$m * a$	number of bites per person per day (product was set to 1, Lindsay and Thomas 2001: 81)
a	h/u
h	proportion of mosquito blood meals taken from people instead of other animals, also expressed as HBI (the model assumes a mean value for h of 0.42 for indoor-resting mosquitoes, Jetten and Takken 1994: 50)
u	length in days of the gonotrophic cycle (interval between each mosquito's blood meal and oviposition)
u	$f_1/(T - g_1)$
f_1	thermal sum, measured in degree days (36.5 at threshold of 9.9 °C, Jetten and Takken 1994: 36)
T	average ambient temperature
g_1	threshold below which development ceases (9.9 °C, Jetten and Takken 1994: 36)
b	proportion of female mosquitoes developing parasites after taking an infected blood meal (as no value could be found in literature for German *Anopheles* mosquitoes, a mean value was calculated out of records from England and the Netherlands. Thus, the assumed value for Germany is 0.14)
p	daily survival probability of an adult female mosquito (the model considers a median value of mortality rate for *Anopheles atroparvus* = 0.029/day_p = 0.97/day, Jetten and Takken 1994: 42)
n	period of parasite development within the adult female mosquitoes in days: sporogonic cycle
n	$f_2/(T - g_2)$
f_2	thermal sum, measured in degree days (105 at threshold of 14.5 °C, Jetten and Takken 1994: 46)
T	average air temperature
g_2	minimum temperature required for parasite development (14.5 °C, Jetten and Takken 1994: 46)

r   rate of recovery of humans from infection from malaria (the usual
       assumption is that the duration of each infection is therefore 1/*r* days.
       According to Lindsay and Thomas (2001: 81), it is assumed that an
       infection would be present for 60 days, resulting in a value for 1/*r* of
       0.0167/day)

The application of the basic reproduction rate formula to the monthly mean
temperature maps (Sect. 2.2.3) enabled detecting potential areas at risk for a tertian
malaria outbreak in Lower Saxony. The resulting maps depict areas with different
transmission periods where tertian malaria could be spread under past, present and
future climate conditions. These operations were performed with the help of the
tool Map Query included in the GIS software ArcView 3.3 distributed by ESRI.

Concerning the prediction maps about areas at risk in future (2020, 2060, 2100),
the same approach was used. To the air temperature grid averaged for the period
1985–2004 the according projected air temperature rise (Sect. 2.2.3) was added for
each map pixel. For each of the future years, two different climate change models
were used: the "best case" represents the lowest predicted temperature values, the
"worst case", by contrast, reflects the highest possible temperature values calculated in the 35 IPCC climate scenarios mentioned above.

2.3 Results

2.3.1 Potential Transmission Gates Calculated
         with Temperature Measurements

The calculations yielded maps on areas at risk of malaria outbreak and maps on the
duration of the potential transmission periods of the tertian malaria pathogen due
to past, recent and future monthly mean air temperatures. The following sections
focus on the potential transmission gates calculated for the periods from 1947 to
1960, 1961 to 1990 and 1985 to 2004 as well as for the projected temperatures in
2020, 2060 and 2100 (Table 2.2).

1947–1960
From 1947 to 1960 in most of the area of Lower Saxony *Anopheles* mosquitoes
can successfully transmit malaria for two month (coverage 81.6 %). The regions
for a three months' spread are by far smaller (16.2 %), but still larger than those of
one month or of no transmission. The longest period of transmission can be found
in the loess belt in the far south of Lower Saxony (Fig. 2.2, left). Smaller spots can
be found along the river Elbe and the far southwest (city of Lingen). Due to
comparable colder weather conditions, the least possible transmission duration
(0–1 month) can be found in the low mountain ranges in the South. The map also
shows those locations where Malaria infections and *Anopheles* findings were

Table 2.2 Number of months of potential malaria transmission expressed in coverage percentage of the investigation area (Lower Saxony) for different climatic periods in past and future (according to Schröder et al. 2007)

No. of months Period (projected temperature rise)	0	1	2	3	4	5	6	Sum (%)
1947–1960	0.9	1.3	81.6	16.2	0.0	0.0	0.0	100.0
1961–1990	2.3	0.9	74.8	21.9	0.0	0.0	0.0	100.0
1985–2004	0.5	0.1	74.4	25.1	0.0	0.0	0.0	100.0
2020 (+0.3 °C)	0.1	0.1	49.9	49.9	0.0	0.0	0.0	100.0
2020 (+0.9 °C)	0.0	0.0	6.0	93.9	0.1	0.0	0.0	100.0
2060 (+0.9 °C)	0.0	0.0	6.0	93.9	0.1	0.0	0.0	100.0
2060 (+3.3 °C)	0.0	0.0	0.0	1.3	4.0	94.7	0.0	100.0
2100 (+1.4 °C)	0.0	0.0	1.9	76.5	21.6	0.0	0.0	100.0
2100 (+5.8 °C)	0.0	0.0	0.0	0.0	0.0	43.9	56.1	100.0

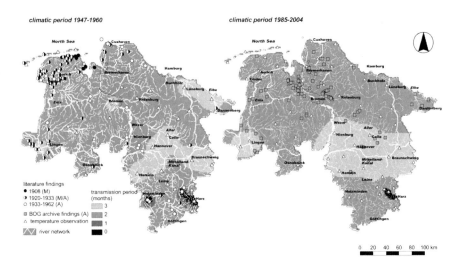

Fig. 2.2 Potential transmission gates of the tertian malaria pathogen for 1947–1960 and 1985–2004 in Lower Saxony (*M* Malaria incidences, *A* Anopheles findings) (according to Schröder et al. 2007)

documented in literature (black/white circles). The main localities can be found at the coastal lowlands south of the North Sea.

1961–1990

From 1961 to 1990, the transmission map reveals a maximum duration of three months for infection, too (coverage 21.9 %). Nevertheless, the predominant transmission duration is two months (74.8 %). Areas where no transmission is possible can be found in the mountain range in the far southeast of Lower Saxony. The districts with a potential *Plasmodium vivax* spread of three months are distributed nearly identically as in the previous period but the spots now become larger.

1985–2004

From 1985 to 2004, the monthly transmission map (Fig. 2.2, right) shows just one very small area where no transmission is possible (Harz Mountains). Most of Lower Saxony is again dominated by a transmission period of two months (74.4 %). Three months areas are located in the centre and again in the south-westmost part of Lower Saxony (25.1 %). The bigger one is a broad belt along the waterway Mittellandkanal that runs from east to west touching the city of Hanover. The second and also smaller spot can be found again in the southwest with the city of Lingen in its centre. The map also shows those locations where *Anopheles* findings were documented in the BOG-Archiv (Sect. 2.2.1) from the early 1980s (grey squares). The main localities can be found at the rivers in the northern third of Lower Saxony.

2.3.2 Potential Transmission Gates Calculated with Estimations of Future Temperatures

The future transmission periods are calculated by the combination of the geosta-tistically computed temperature maps and selected IPCC climate scenarios. Accordingly, the best case would be the lowest expected temperature rise for the respective year, the worst case the highest prospected temperature rise.

Year 2020

The best case scenario 2020 was based on a predicted climate warming of 0.3 °C. The areas where no transmission of tertian malaria is possible together with the areas of one month duration represent less than one percentage of Lower Saxony (Fig. 2.3, left). The rest of the federal state refers equally to transmission areas of two and three months. While the first is located in the northern and southern parts, the second is a central broad belt from the west to the east of Lower Saxony.

In the worst case expecting an air temperature rise of 0.9 °C, for the first time the disease might be transmitted for four months (0.1 %). Also for the first time, whole Lower Saxony is a potential transmission country without any exception, regions with a potential transmission gate of three months are most frequent (coverage 93.9 %). Only four regions (6.0 %) remain, representing zones with a transmission period of two months (low mountain range in the far south and spots near the cities of Hamburg and Aurich in the north).

Year 2060

The best case-prediction 2060 is exactly the same as the 2020 worst case, i.e. the temperature rise is expected to be 0.9 °C. The IPCC presumes a worst case temperature rise of 3.3 °C. Thus, the transmission map of that year reveals that Lower Saxony is almost totally situated with five months of transmission

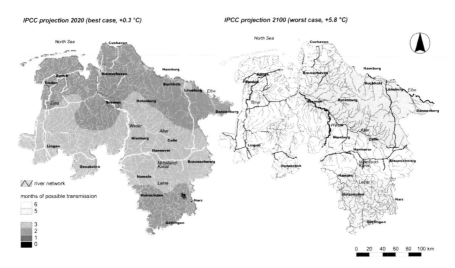

Fig. 2.3 Potential transmission gates of the tertian malaria pathogen projected for 2020, best case (T + 0.3 °C), and 2100, worst case (T + 5.8 °C) in Lower Saxony (according to Schmidt et al. 2008)

(coverage 94.7 %). Again, only the highlands in the far south and the coastal zone in the north make an exception in duration (3–4 months).

Year 2100

The best case scenario estimates a minimum climate warming of 1.4 °C and includes September as the fourth month of a successful transmission of malaria. The main transmission period is three months (coverage 76.5 %). Regions with four months of transmission (21.6 %) are located in a broad belt along the waterway Mittellandkanal in the south and around the city of Lingen in the southwest most part of Lower Saxony. Now, also coastal areas at the North Sea are included and tidal mud flat islands, too. The worst case calculation is based on a temperature rise of +5.8 °C (Fig. 2.3, right). Given this, Lower Saxony is divided into two parts of transmission length: The western part reveals a predicted transmission period of six months (coverage 56.1 %). These areas represent those regions which are usually warmer in comparison to the rest of the country. A disease spread of five months (43.9 %) can be found within the rectangle of the river Weser in the northwest and the waterway Mittellandkanal in the south. Another area covers the low mountain range in the south most part of Lower Saxony.

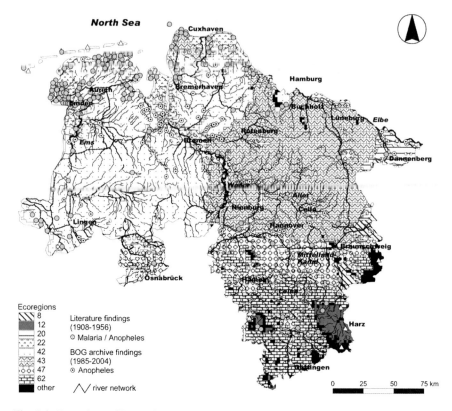

Fig. 2.4 Ecoregions of Lower Saxony calculated by CART (Classification and regression trees) from the data in Table 2.1 and briefly described in Table 2.3 (according to Schröder and Schmidt 2008)

2.3.3 Ecoregionalised Transmission Gates

The ecological land classification calculated on the data described in Table 2.1 by use of Classification and regression trees (CART; Breiman et al. 1984) yielded a hierarchy of maps differentiating Germany by an according number of ecoregions. Using the map with 21 ecologically defined regions in Germany (Fig. 2.1), 8 of them are located in Lower Saxony, each of the them covering more than 1 % of Lower Saxony's territory each (Fig. 2.4) (Schröder 2006; Schröder et al. 2006). Each of these ecoregions is detailed by descriptive statistics of selected characteristics such as climate, soil texture, altitude above sea level, and Potential natural vegetation (PNV), giving a portrayal of the natural floristic biodiversity and indicating areas with similar site properties and natural growth properties (Bohn et al. 2000/2003) (Table 2.3).

Table 2.3 Relevant characteristics of the ecoregions of Lower Saxony (according to Schröder and Schmidt 2008)

Landscape characteristics		12	20	22	42	43	47	62	8
Frequency (Germany) (%)		6.3	3.8	1.6	7.3	7.2	8.1	14.0	4.1
PNV[2]		F60	F71	U21	F5	F52	F68	F55	F54
		F59	F68		F52	F5	U7	F70	U19
		F61	F52		F53		F52		F68
		F56					F23		F52
Soil texture[3]		Loamy sand, gritty	Loamy sand, loam	Sandy clay silt	Sand, loamy sand	Sand, silty sand	Clay silt to silty loam	Sandy clay loam	Loamy sand, loam
		Sandy clay, loam	Sand, silty sand		Peat	Loamy sand, loam	Clayey silt to loamy clay	Sand to silty loam	Sand, loamy sand
		Sand to silty loam	Sand, loamy sand		Loamy sand, loam	Sand, loamy sand	Loamy sand,silty loam	Clay silt to silty loam	Peat
			Peat		Sand, silty sand	Peat			Sand, silty sand
Altitude above sea level (m)	Min	158.0	0.0	0.0	0.0	0.0	0.0	156.0	0.0
	Med	598.0	50.0	2.0	35.0	38.0	90.0	282.0	45.0
	Max	1414.0	155.0	20.0	154.0	155.0	155.0	673.0	152.0
Global radiation (Wh/month)	Min	65.0	62.9	59.0	61.6	58.0	61.6	61.4	61.8
	Med	334.4	335.3	326.7	327.3	328.7	330.4	333.2	334.9
	Max	554.8	566.7	533.3	516.7	553.3	554.8	558.1	556.7
Precipitation (mm/month)	Min	37.8	22.0	32.3	37.8	28.3	27.3	30.8	24.75
	Med	87.9	48.3	65.8	63.4	59.0	60.9	68.4	45.6
	Max	214.5	83.0	104.0	98.5	107.8	140.2	172.0	78.3
Evaporation rate (mm/month)	Min	6.0	5.0	5.0	6.0	5.0	6.0	8.0	5.0
	Med	40.5	48.0	44.9	46.5	46.7	48.0	47.1	49.0
	Max	91.0	104.0	92.0	91.8	106.0	107.0	108.5	99.0

(continued)

Table 2.3 (continued)

Landscape characteristics		Ecoregions[1]							
		12	20	22	42	43	47	62	8
Air temperature (°C/ month)	Min	−5.0	−1.6	0.2	−0.2	−1.2	−1.4	−0.7	−1.5
	Med	6.5	8.2	8.6	8.9	8.4	9.3	8.3	8.5
	Max	17.7	17.7	17.3	18.4	19.3	19.7	19.5	18.7

[1] Covering at least 1 % of the territory

[2] Contributing at least 10 % to the respective ecoregion, after Bohn et al. (2000/2003), see Annex Table 2.3

[3] Contributing at least 10 % to the respective ecoregion

Min Minimum, *Med* Median, *Max* Maximum

Annex Table 2.3

Code	Potential natural vegetation according to Bohn et al. (2000/2003)
F 5	East Armorican thermophilous oak forests (*Quercus petraea, Q. robur*, locally *Quercus pyrenaica*) of the Loire-basin with *Peucedanum gallicum, Festuca heterophylla, Luzula forsteri, Erica scoparia*
F 23	Galician-north Lusitanian oak forests (*Quercus robur, Q. pyrenaica*) with *Betula pubescens subsp. celtiberica, Cytisus striatus, Dryopteris aemula, Anemone trifolia subsp. albida, Omphalodes nitida*
F 52	South subatlantic sessile oak-hornbeam forests (*Carpinus betulus, Quercus petraea*), rich in thermophilous species, alternating with thermophilous mixed beech and oak forests (*Fagus sylvatica, Quercus petraea, Ulmus minor, Carpinus betulus, Tilia cordata*, partly *Quercus pubescens*)
F 53	South subatlantic thermophilous sessile oak-hornbeam forests (*Carpinus betulus, Quercus petraea*) with *Quercus pubescens*, partly with *Buxus sempervirens*
F 54	Sessile oak-hornbeam forests (*Carpinus betulus, Quercus petraea*) on siliceous soils with *Castanea sativa* in the west Alps
F 55	Central European sessile oak-hornbeam forests (*Carpinus betulus, Quercus petraea*), mostly with *Fagus sylvatica*, with *Sorbus torminalis, Carex montana, Hepatica nobilis*
F 56	Pre-Sudeten sessile oak-hornbeam forests (*Carpinus betulus, Quercus petraea, Q. robur*), mostly with *Fagus sylvatica*, submontane type with *Senecio ovatus, Prenanthes purpurea*, partly *Carex brizoides*
F 59	Northwest *Pannonian thermophilous* oak-hornbeam forests (*Carpinus betulus, Quercus petraea, Q. robur*) with *Primula veris, Carex michelii, Pulmonaria mollis*
F 60	Transsylvanian sessile oak-hornbeam forests (*Carpinus betulus, Quercus petraea*) with *Lathyrus hallersteinii*
F 61	Northeast pre-Carpathian sessile oak-hornbeam forests (*Carpinus betulus, Quercus petraea, Q. robur*) with *Fagus sylvatica*, with *Aposeris foetida*
F 68	East Moesian-west Pontic mixed sessile oak-hornbeam-silver lime forests (*Tilia tomentosa, Carpinus betulus, Quercus petraea, Q. dalechampii, Q. polycarpa*), partly with *Carpinus orientalis*, with *Nectaroscordum siculum subsp. bulgaricum*
F 70	Baltic-south Sarmatian lime-pedunculate oak forests (*Quercus robur, Tilia cordata*), partly with *Picea abies*
F 71	North Ukrainian-south Sarmatian lime-pedunculate oak forests (*Quercus robur, Tilia cordata*) with *Fraxinus excelsior, Acer campestre, A. tataricum*
U 7	Irish-British hardwood alluvial forests (*Quercus robur, Fraxinus excelsior, Salix atrocinerea*) in combination with willow alluvial forests (*Salix fragilis, S. alba, S. viminalis*)
U 19	Danubian hardwood alluvial forests (*Fraxinus angustifolia subsp. danubialis, F. pallisae, Quercus robur, Q. pedunculiflora, Ulmus minor*) in combination with poplar and willow alluvial forests (*Populus alba, P. nigra, Salix alba*)
U 21	Caucasian hardwood alluvial forests (*Quercus robur, Ulmus minor*) in combination with poplar and willow alluvial forests (*Populus x canescens, P. alba, Salix alba*) and willow scrub (*Salix triandra*)

The geostatistical estimation resulted in air temperature maps (Sect. 2.2.3) which were used for the application of the R_0 formula. In result, maps on areas at risk of malaria outbreak as well as maps on the length of the transmission gates of the tertian malaria pathogen were derived considering recent (Sect. 2.3.1) and future monthly mean air temperatures (Sect. 2.3.2). In a next step, each of the risk maps was spatially differentiated in terms of natural landscapes by intersection with the map on ecoregions in a GIS (Sect. 2.2.4). Those ecoregions covering more than 1 % of Lower Saxony's territory are described in Table 2.3 with regard to those landscape characteristics which were used for ecoregionalisation (Table 2.1). The following paragraphs focus on the seasonal transmission gates within these ecoregions considering the periods 1947–1960, 1961–1990, and 1985–2004 as well as the projected future development in 2020, 2060 and 2100 (Tables 2.4–2.11).

From 1947 to 1960, areas with potential malaria transmission over two months per year cover 81.7 % of the territory of Lower Saxony. A three-month transmission gate would have been possible on 16.1 % of the land area. These two transmission regions are mainly distributed in ecoregions (ER) 42, 43 (North German coastal heath land), and 47 (Table 2.4). The remaining area is situated by a seasonal transmission gate of only one month or no transmission is possible. The longest seasonal transmission gate can be found in the loess belt in the far south of Lower Saxony (ER 47) with mean annual air temperatures of 9.3 °C. Smaller spots can be found along the river Elbe and the far south-west (ER 43). Due to cold weather conditions (mean annual temperatures 6.5–8.3 °C), the least possible annual transmission gate (0–1 month) can be found in the low mountain ranges in the south (ER 12 and ER 62) with average altitudes between 282 m and 598 m above sea level (see Table 2.3).

From 1961 to 1990, malaria transmission would have been possible over three months (21.8 %), whereas the predominant annual transmission period accounts for two months (74.9 %) (Table 2.5). Again, areas where no transmission was possible could be found in the mountain range in the far south-east of Lower Saxony (ER 12 and ER 62). The regions with a possible *P. vivax* spread of three months were almost identical to those in the previous period but the spots became larger.

From 1985 to 2004, just one very small area could be identified where no transmission was possible (ER 12, Harz Mountains) (Table 2.6). The rest of

Table 2.4 Potential malaria seasonal transmission gate (months) in the ecoregions of Lower Saxony 1947–1960 (percentage per ecoregion; total = percentage of overall coverage) (according to Schröder and Schmidt 2008)

Months	Ecoregions (ER)[*] after Schröder et al. (2006)								Total (%)
	12 (%)	20 (%)	22 (%)	42 (%)	43 (%)	47 (%)	62 (%)	8 (%)	
0	81.1	0.0	0.0	0.0	0.0	0.0	3.6	0.0	0.9
1	46.2	0.0	0.0	0.0	0.0	0.6	26.3	0.0	1.3
2	0.2	1.3	8.4	41.9	24.6	11.8	8.4	0.9	81.7
3	0.0	6.2	0.9	5.8	39.4	32.6	7.1	3.2	16.1
Total 1947–1960	1.5	2.1	7.0	35.2	26.4	14.9	8.4	1.3	100.0

[*] Covering at least 1 % of the territory of Lower Saxony

Table 2.5 Potential malaria seasonal transmission gate (months) in the ecoregions of Lower Saxony 1961–1990 (percentage per ecoregion; total = percentage of overall coverage) (according to Schröder and Schmidt 2008)

Months	Ecoregions (ER)[*] after Schröder et al. (2006)								Total (%)
	12 (%)	20 (%)	22 (%)	42 (%)	43 (%)	47 (%)	62 (%)	8 (%)	
0	50.5	0.0	0.0	0.0	0.0	7.3	21.8	0.0	2.3
1	14.0	0.0	0.0	7.9	3.5	7.9	38.6	0.0	1.0
2	0.2	0.9	9.3	43.9	22.4	12.2	8.0	0.4	74.9
3	0.0	6.4	0.0	10.1	44.3	25.3	6.9	4.4	21.8
Total 1961–1990	1.5	2.1	7.0	35.2	26.5	14.9	8.4	1.3	100.0

[*] Covering at least 1 % of the territory of Lower Saxony

Table 2.6 Potential malaria seasonal transmission gate (months) in the ecoregions of Lower Saxony 1985–2004 (percentage per ecoregion; total = percentage of overall coverage) (according to Schröder and Schmidt 2008)

Months	Ecoregions (ER)[*] after Schröder et al. (2006)								Total (%)
	12 (%)	20 (%)	22 (%)	42 (%)	43 (%)	47 (%)	62 (%)	8 (%)	
0	98.1	0.0	0.0	0.0	0.0	0.0	0.0	0.0	0.4
1	92.3	0.0	0.0	0.0	0.0	0.0	0.0	0.0	0.1
2	1.3	1.8	9.4	42.8	24.5	8.2	8.6	0.2	74.5
3	0.0	2.8	0.0	13.0	32.9	35.3	7.7	4.7	25.0
Total 1985–2004	1.5	2.1	7.0	35.2	26.4	14.9	8.4	1.3	100.0

[*] Covering at least 1 % of the territory of Lower Saxony

Lower Saxony was characterised by a potential transmission gate of two months (74.5 %). Areas with a three month transmission potential are located in the middle and in the south-western part of Lower Saxony (25 %) (ER 43 and ER 47).

The *best case scenario* 2020 is based on a predicted temperature increase of about 0.3 °C when compared to the reference period 1961–1990. The areas where no transmission of tertian malaria would be possible together with the areas of one month transmission represent less than 1 % of the territory of Lower Saxony (Table 2.7). The remainder of the country was divided equally in transmission gates of two and three months. While the former was located in the northern and southern parts, the latter formed a central broad belt from western to eastern Lower Saxony.

In the 2020 *worst case scenario*, expecting an average air temperature rise of 0.9 °C referred to the reference period 1961–1990, malaria might be transmittable for four months (0.1 % of the land area) for the first time (Table 2.8). Additionally, in the whole territory of Lower Saxony, transmission was expected to be possible and a transmission gate of three months prevails (93.9 %). Only four ecoregions (6.0 %) may remain, representing zones with a transmission gate of two months (the low mountain range in the south, ER 12 and ER 62, and spots in the north, ER 42 and ER 43).

Table 2.7 Potential malaria seasonal transmission gate (months) in the ecoregions of Lower Saxony 2020 best case (b.c.) (percentage per ecoregion; total = percentage of overall coverage) (according to Schröder and Schmidt 2008)

Months	Ecoregions (ER)[*] after Schröder et al. (2006)								Total (%)
	12 (%)	20 (%)	22 (%)	42 (%)	43 (%)	47 (%)	62 (%)	8 (%)	
0	100.0	0.0	0.0	0.0	0.0	0.0	0.0	0.0	0.1
1	100.0	0.0	0.0	0.0	0.0	0.0	0.0	0.0	0.2
2	2.5	2.6	13.4	36.4	23.9	6.4	10.4	0.2	50.1
3	0.0	1.6	0.5	34.1	29.2	23.6	6.3	2.4	49.7
Total 2020 b.c.	1.5	2.1	7.0	35.2	26.4	14.9	8.4	1.3	100.0

[*] Covering at least 1 % of the territory of Lower Saxony

Table 2.8 Potential malaria seasonal transmission gate (months) in the ecoregions of Lower Saxony 2020 worst case (w.c.)/2060 best case (b.c.) (percentage per ecoregion; total = percentage of overall coverage) (according to Schröder and Schmidt 2008)

Months	Ecoregions (ER)[*] after Schröder et al. (2006)								Total (%)
	12 (%)	20 (%)	22 (%)	42 (%)	43 (%)	47 (%)	62 (%)	8(%)	
2	24.0	0.0	6.6	12.9	14.3	3.8	24.2	0.0	6.0
3	0.1	2.2	7.0	36.6	27.2	15.6	7.4	1.4	93.9
Total 2020 w.c./2060 b.c.	1.5	2.1	7.0	35.2	26.4	14.9	8.4	1.3	100.0

[*] Covering at least 1 % of the territory of Lower Saxony

The 2060 *best case scenario* equals the 2020 worst case, i.e., the average temperature rise compared to the reference period 1961–1990 is expected to be 0.9 °C. The IPCC presumes an average worst case temperature rise of 3.3 °C. Thus, a transmission gate of five months dominated in Lower Saxony (coverage 94.3 %) (Table 2.9). Again, only the highlands in the south (ER 12 and ER 62) and the coastal zone in the north (ER 22) differed (3–4 months).

The 2100 *best case scenario* estimates a minimum climate warming of 1.4 °C when compared to the reference period 1961–1990 and included September as the fourth month susceptable for transmission of malaria (Table 2.10). The most frequent seasonal transmission gate lasted three months (76.6 %). Regions with four months transmission (21.4 %) were located in a broad belt in the south and in the south-western part of Lower Saxony (ER 42 and ER 47). Also coastal areas at the North Sea were included and tidal mud flat islands, too (ER 22).

The worst case calculation was based on a temperature rise of 5.8 °C compared to the reference period 1961–1990. Lower Saxony would then be divided into two parts with differing transmission gates (Table 2.11). The western part revealed a predicted transmission gate of six months (coverage 55.8 %). These areas represent those regions which are usually warmer compared with the other parts of the country (ER 42) showing annual mean temperatures of 8.9 °C (1961–1990), whereas the average for Lower Saxony was 8.6 °C in that period. A seasonal

Table 2.9 Potential malaria seasonal transmission gate (months) in the ecoregions of Lower Saxony 2060 worst case (w.c.) (percentage per ecoregion; total = percentage of overall coverage) (according to Schröder and Schmidt 2008)

Months	Ecoregions (ER)[*] after Schröder et al. (2006)								Total (%)
	12 (%)	20 (%)	22 (%)	42 (%)	43 (%)	47 (%)	62 (%)	8 (%)	
3	82.8	0.0	0.0	0.0	0.0	0.0	3.4	0.0	1.2
4	7.3	0.0	17.7	8.2	0.0	4.7	41.9	0.0	4.5
5	0.1	2.2	6.6	36.9	28.1	15.6	6.8	1.4	94.3
Total 2060 w.c.	1.4	2.1	7.0	35.2	26.5	14.9	8.4	1.3	100.0

[*] Covering at least 1 % of the territory of Lower Saxony

Table 2.10 Potential malaria seasonal transmission gate (months) in the ecoregions (12–8) of Lower Saxony 2100 best case (b.c.) (percentage per ecoregion; total = percentage of overall coverage) (according to Schröder and Schmidt 2008)

Months	Ecoregions (ER)[*] after Schröder et al. (2006)								Total (%)
	12 (%)	20 (%)	22 (%)	42 (%)	43 (%)	47 (%)	62 (%)	8 (%)	
2	65.5	0.0	0.0	2.2	1.7	0.0	9.9	0.0	1.9
3	0.3	2.2	6.5	36.7	31.4	10.9	9.1	0.7	76.6
4	0.0	1.8	9.4	32.7	10.9	30.5	5.6	3.5	21.4
Total 2100 b.c.	1.5	2.1	7.0	35.2	26.4	14.9	8.4	1.3	100.0

[*] Covering at least 1 % of the territory of Lower Saxony

Table 2.11 Potential malaria seasonal transmission gate (months) in the ecoregions (12–8) of Lower Saxony 2100 worst case (w.c.) (percentage per ecoregion; total = percentage of overall coverage) (according to Schröder and Schmidt 2008)

Months	Ecoregions (ER)[*] after Schröder et al. (2006)								Total (%)
	12 (%)	20 (%)	22 (%)	42 (%)	43 (%)	47 (%)	62 (%)	8 (%)	
5	3.4	3.3	2.0	13.7	46.0	11.2	15.1	1.2	44.2
6	0.0	1.1	10.9	52.1	10.9	17.8	3.0	1.4	55.8
Total 2100 w.c.	1.5	2.1	7.0	35.2	26.4	14.9	8.4	1.3	100.0

[*] Covering at least 1 % of the territory of Lower Saxony

transmission gate of five months (coverage 44.2 %) was found within the rectangle of the river Weser in the north-west and the waterway 'Mittellandkanal' in the south (ER 43). Another area covered the low mountain range in the south of Lower Saxony (ER 12).

2.3.4 Discussion

The investigation revealed areas in Lower Saxony that are susceptible for a new tertian malaria outbreak transmitted by *A. atroparvus* due to their high summer mean temperatures. The calculations corroborate that although malaria is not an endemic disease in Germany, the risk of an autochthonous transmission does exist (Krüger et al. 2001). Pathogens of the respective malaria types enter the country by infected travellers and by mosquitoes from endemic areas.

Concerning the period from 1947 to 2004, in terms of transmission length, the dominant transmission period was two months (74–82 %). In all the three periods, the regions with the highest risk of a disease spread are located around the city of Lingen in the southwest and along a broad belt in the south touching the cities of Braunschweig, Hanover and Nienburg. A similarity between former malaria zones and recent *Anopheles* findings listed in the BOG-Archiv can be noticed. Some areas mentioned there are exactly those which represent the highest risk of a *Plasmodium vivax* transmission for all investigated periods. In the southern parts of Germany, *Anopheles* findings had not been determined to the species level. However, recent studies give evidence that *Anopheles atroparvus* and *Anopheles messeae* are still present in the coastal areas (Wilke et al. 2006). It is just this area, where best case scenarios already calculated a four months risk of transmission in 2060. And it is this area, where malaria tertiana occurred up to the 1950s (Weyer 1956).

Furthermore, it was tested to what extend the predicted climate warming prolongs the transmission period of the tertian malaria pathogen. Hence, the formula of the model was applied to IPCC predicted climate warming scenarios. Outcomes were calculated for the years 2020, 2060 and 2100, illustrated by maps of best and worst case scenarios at each year. In worst case, the predicted maximum air temperature rise for the present century is expected to be more than four times higher than in the best case scenario. As the difference between the chosen years is, in average, 2.45 °C for the worst case, the mean temperature increase is 0.55 °C in best case considering the development of the 80 years' period between 2020 and 2100. By comparing all the maps, some striking trends become obvious. First of all, the climate warming between 2020 and 2060 prolongs the transmission period for nearly all areas of Lower Saxony by two months. This does not hold true for the last prospected period (2100), where an average prolongation of one another month is expected. The second point is similar to one mentioned when discussing the best case scenarios. The coastal areas show different warming characteristics than the continental regions. Up from 2020 to 2060 and 2100 at last, the potential tertian malaria transmission gate turns from two months to four and then directly to six. As explained before, this climate condition is a result of the warm west winds from the North Sea bringing mild temperature conditions in autumn. This development gets consequently obvious, when months like September or October become transmission months of tertian malaria, like they do in terms of the worst case maps.

References

Bartels F (1997) Ein Fuzzy-Auswertungs- und Krigingsystem für raumbezogene Daten. Diploma thesis, Universität Kiel

Bohn U, Neuhäusl R, Gollub G, Hettwer C, Nehäuslová Z, Schlüter H, Weber H (2000/2003) Map of the Natural Vegetation of Europe. Scale 1:2.5 million. Part 1: explanatory text with

CD-ROM. Part 2: legend. Part 3: maps (9 Sheets 1:2.5 million, Legend Sheet, General Map 1:10 million). Landwirtschaftlicher Verlag, Münster

Breiman L, Friedman JH, Olshen RA, Stone CJ (1984) Classification and regression trees. Wadsworth International Group, Belmont

Eritja R, Aranda C, Padrós J, Goula M, Lucientes J, Escosa R, Marquès E, Cáceres F (2000) An annotated checklist and bibliography of the mosquitoes of Spain (Diptera: Culicidae). Eur Mosq Bull 8:11–42

Hackett LW, Missiroli A (1935) The varieties of Anopheles maculipennis and their relation to the distribution of malaria in Europe. Riv Malariologia 14(1):1

Heinz HJ (1950) Neuere Untersuchungen über die Verbreitung von Anopheles maculipennis in Hamburg. J Appl Entomol 31:304–333

Intergovernmental Panel of Climate Change (IPCC) (2001) Climate change. The scientific basis. Cambridge University Press, Cambridge

Jetten TH, Takken W (1994) Anophelism without malaria. Wageningen Agricultural University Papers 94 (5)

Krüger A, Rech A, Su XZ, Tannich E (2001) Two cases of autochthonous *Plasmodium falciparum* malaria in Germany with evidence for local transmission by indigenous *Anopheles plumbeus*. Trop Med Int Help 6:983–985

Kubica-Biernat B (1999) Distribution of mosquitoes (Diptera: Culicidae) in Poland. Eur Mosq Bull 5:1–17

Lindsay SW, Thomas CJ (2001) Global warming and risk of vivax malaria in Great Britain. Glob Change Hum Health 2(1):80–84

Maier WA, Grunewald J, Habedank B, Hartelt K, Kampen H, Kimmig P, Naucke T, Oehme R, Vollmer A, Schöler A, Schmitt C (2003) Mögliche Auswirkungen von Klimaveränderung auf die Ausbreitung von primär humanmedizinisch relevanten Krankheitserregern über tierische Vektoren sowie auf die wichtigen Humanparasiten in Deutschland. Climate Change 05/03. Umweltbundesamt, Berlin

Martens P, Kovats RS, Nijhof S, de Vries P, Livermore MTJ, Bradley DJ, Cox J, McMichael AJ (1999) Climate change and future population at risk of malaria. Glob Environ Change 9:89–107

Martini E (1920a) *Anopheles* in der näheren und weiteren Umgebung von Hamburg und ihre voraussichtliche Bedeutung für die Volksgesundheit. Abhandlungen aus dem Gebiet der Naturwissenschaften 21(2)

Martini E (1920b) Anopheles in Niedersachsen und die Malariagefahr. Hyg Rundsch 22:673–677

Martini E (1946) Lehrbuch der medizinischen Entomologie. Gustav Fischer, Jena

Mühlens P (1930) Malaria. Neue Deutsche Klinik. Handwörterbuch der Praktischen Medizin mit besonderer Berücksichtigung der Inneren Medizin, der Kinderheilkunde und ihrer Grenzgebiete 7(31):122–149

Piotrowski JA, Bartels F, Salski A, Schmidt G (1996) Geostatistical regionalization of glacial aquitard thickness in northwestern Germany, based on fuzzy kriging. Math Geol 28(4):437–452

Ramsdale C, Snow K (2000) Distribution of the genus *Anopheles* in Europe. Eur Mosq Bull 7:1–26

Reiter P (2008) Global warming and malaria: knowing the horse before hitching the cart. Malar J 7(1):S3

Romi R, Pierdominici G, Severini C, Tamburo A, Cocchi M, Menichetti D, Pili E, Marchi A (1997) Status of malaria vectors in Italy. J Med Entomol 34:263–271

Ross R (1911) The prevention of malaria. John Murray, London

Schaffner F (1998) A revised checklist of French mosquitoes. Eur Mosq Bull 2:1–9

Schmidt G (2002) Eine multivariat-statistisch abgeleitete ökologische Raumgliederung für Deutschland. dissertation.de, Berlin

Schmidt G, Holy M, Schröder W (2008) Vector-associated diseases in the context of climate change: analysis and evaluation of the differences in the potential spread of tertian malaria in the ecoregions of Lower Saxony. Ital J Public Health 5(4):243–250

Schröder W (2006) GIS, geostatistics, metadata banking, and tree-based models for data analysis and mapping in environmental monitoring and epidemiology. Int J Med Microbiol 296(40):23–36

Schröder W, Schmidt G (2001) Defining ecoregions as framework for the assessment of ecological monitoring networks in Germany by means of GIS and classification and regression trees (Cart). Gate to EHS 2001:1–9

Schröder W, Schmidt G (2008) Mapping the potential temperature-dependent tertian malaria transmission within the ecoregions of Lower Saxony (Germany). Int J Med Microbiol 298(S1):38–49

Schröder W, Schmidt G, Hasenclever J (2005) Bioindication of climate change by means of mapping plant phenology on a regional scale. A geostatistically based correlation analysis of data on air temperature and phenology by the example of Baden-Württemberg (Germany). Environ Monit Assess 130:27–43

Schröder W, Schmidt G, Hornsmann I (2006) Landschaftsökologische Raumgliederung Deutschlands. In: Fränzle O, Müller F, Schröder W (eds) Handbuch der Umweltwissenschaften. Grundlagen und Anwendungen der Ökosystemforschung. ecomed, München, Kap. V-1.9, 17. Erg.Lfg.:1–100

Schröder W, Schmidt G, Bast H, Pesch R, Kiel E (2007) Pilot-study on GIS-based risk modelling of a climate warming induced tertian malaria outbreak in Lower Saxony (Germany). Environ Monit Assess 133:483–493

Smith DL, McKenzie FE (2004) Statics and dynamics of malaria infection in Anopheles mosquitoes. Malar J 3:13

Snow RW, Ikoku A, Omumbo J, Ouma J (1999) The epidemiology, politics and control of malaria epidemics in Kenya: 1900–1998. Roll Back Malaria. Resource network on epidemics. World Health Organisation, Nairobi

Webster R, Oliver MA (2001) Geostatistics for environmental scientists. Wiley, Chichester

Weyer F (1940) Malaria und Malariaübertragung in Ostfriesland. Archiv für Schiffs- und Tropenmed 44:1–73

Weyer F (1956) Bemerkungen zum Erlöschen der ostfriesischen Malaria und zur Anopheles-Lage in Deutschland. Z Tropenmed Parasitol 7:219–228

Wilke A, Kiel E, Schröder W, Kampen H (2006) Anophelinae (Diptera: Culicidae) in ausgewählten Marschgebieten Niedersachsens: Bestandserfassung, Habitatbindung und Interpolation. Mitt Dtsch Ges Allg Angew Entomol 15:357–362

Chapter 3
Case Study 2: Modelling Potential Risks of Malaria Outbreaks in Germany

Abstract As shown in *case study* 1 *Lower Saxony*, climate warming can change the geographic distribution and intensity of the transmission of vector-borne diseases such as malaria. The transmitted parasites usually benefit from increased temperatures, as both their reproduction and development are accelerated. Lower Saxony has been a malaria region until the 1950s and the vector species are still present throughout Germany. This gave reason to investigate whether an autochthonous transmission could take place across whole Germany and not only in Lower Saxony if the malaria pathogen is introduced again. As introduced in *case study* 1, in *case study* 2 as well the spatial distribution of potential temperature-driven malaria transmission was investigated using the basic reproduction rate (R_0) to model and geostatistically map areas at risk of an outbreak of tertian malaria based on measured (1961–1990, 1991–2007) and predicted (1991–2020, 2021–2050, 2051–2080) monthly mean air temperature data. The computations performed in a GIS environment produced maps showing that during the period 1961–1990, the seasonal transmission gate in Germany ranges from 0 to 4 months and then expands up to 5 months in the period 1991–2007. In contrast to the calculations performed for Lower Saxony, there was an improved database available considering the future development of air temperatures in Germany. For the projection of future trends, the regional climate models REMO and WettReg were used each with two different emission scenarios (A1B, B1) spatially differentiating the increase in air temperatures. Both modelling approaches resulted in prolonged seasonal transmission gates in the future enabling malaria transmissions during up to 6 months in the climate reference period 2051–2080 (REMO, scenario A1B). The modelling presented in this study can help to identify areas at risk and initiate prevention. The findings of *case study* 2 could also help to investigate the spread of related diseases caused by temperature-dependent pathogens and vectors, including those being dangerous for livestock as well, e.g. bluetongue disease. Without consideration of other relevant influences besides air temperature, one can assume that in case of a reintroduction and a permanent, autochthonous establishment of *Plasmodium vivax*, the potential of a return of tertian malaria in Germany is possible. Yet, so far, the number of imported malaria cases in Germany is assessed to be too low for a re-establishment of autochthonous

W. Schröder and G. Schmidt, *Modelling Potential Malaria Spread in Germany by Use of Climate Change Projections*, SpringerBriefs in Environmental Science, DOI: 10.1007/978-3-319-03823-0_3, © The Author(s) 2014

malaria. Contrary to Germany, other European countries such as the UK and Italy have already undertaken studies in the field of malaria risk assessment. Germany should also face up to this situation and intensify its research activities. The focus should thereby not only be directed on humans but also on livestock, especially for those regions where cattle or poultry are kept in high densities, like Lower Saxony in northern Germany. For a more accurate spatiotemporal analysis of the risk potential, further influencing factors should be considered, e.g. by including maps on the distribution of natural and artificial water bodies (wetlands and river networks), precipitation, humidity, as well as on population exposure and livestock density.

Keywords *Anopheles atroparvus* · Basic reproduction rate · Climate projections · Malaria tertiana · Vector-borne diseases

3.1 Background, Aim and Scope

The global rise in air temperature amounted to 0.74 °C in the 100-year period 1906–2005. Further, worldwide all the 10 warmest years between 1880 and 2007 have occurred since 1998 according to temperature data recording.[1] Depending on the assumed scenarios concerning the development of greenhouse gas emissions and the increase in the world's population, the global mean temperature is projected to rise up to 6.4 °C till 2099 (IPCC 2007). Consequently, the role of climate change in the transmission of vector-borne diseases is subject to numerous investigations dealing with, e.g. dengue fever (Tseng et al. 2009), tick-borne diseases (Hartelt et al. 2008) and especially malaria (Zhang et al. 2008). The latter is the most widespread and most important vector-borne disease worldwide with more than 3 billion people (46 %) living under threat (WHO 2005). Due to the observed and further predicted rise in temperature, vector species can establish at higher latitudes and altitudes. For the species *Anopheles farauti*, living in the southern hemisphere, Bryan et al. (1996) projected that they will appear up to 800 km southward by 2030. In this context, not only new areas are of importance but also such regions having been endemic in former times and where the diseases became extinct for various reasons. This is the case for Germany where malaria tertiana was endemic in the north-western parts of the country till the 1950s. Due to the application of DDT, the drainage of wetlands and improved medical care the disease finally became eradicated. But as only the malaria pathogen *Plasmodium vivax* vanished, the vector itself is still present throughout Germany (Dalitz 2005; Maier et al. 2003), meaning that a new onset of the disease is possible.

Accordingly, *case study* 2 models the potential spread of tertian malaria in Germany which could occur in case of a reintroduction of the causative malaria

---

[1] http://www.statista.com/statistics/158082/climate-change-the-10-warmest-years-since-1880.

pathogen (Holy et al. 2011). To this end, the development of *P. vivax* in *Anopheles atroparvus* was modelled by means of the basic reproduction rate (R_0) (Sect. 2.2.5) and mapped on the basis of measured and projected data on mean monthly air temperatures (Sect. 3.2.3). In contrast to *case study* 1 covering the German federal state Lower Saxony only, in *case study* 2 the potential spread of the disease was evaluated nationwide, and, for the first time, air temperature projections derived from two climate models were used to estimate the potential future malaria transmission (Sect. 3.2.2).

3.2 Materials and Methods

3.2.1 Basic Information on Anopheles Mosquitoes

Modelling and mapping of the potential temperature-dependent malaria spread requires quantitative information on the ontology and ecology of the relevant Anopheles vector. From the literature research (Sect. 2.2.1) we learned that *Anopheles* (Diptera, Culicidae) is one of 41 mosquito genera. The females of 30–40 of the 430 Anopheles species transmit the malaria pathogen, Plasmodium, to humans. Six of the 16 European Anopheles species occur in Germany, three of which were responsible for malaria transmissions till the post-war era (*An. atroparvus, An. maculipennis, An. messeae*) (Ramsdale and Snow 2000). *An. atroparvus* transmits *P. vivax*, the causative agent of tertian malaria, which is also called vivax malaria and commonly known as 'marsh fever' in the coastal regions of northern Germany (Hackett and Missiroli 1935; Martini 1920; Mühlens 1930; Weyer 1940). The mosquitoes need 105 degree days with temperatures $\geq$14.5 °C to become infectious (Jetten and Takken 1994). The temperature threshold values needed for the development of the aquatic stages of *An. atroparvus* were adopted from the literature (Jetten and Takken 1994). In the climate of the first half of the twentieth century, two to three generations of Anopheles grew up per annum in northern Germany (Heinz 1950; Martini 1946). In extraordinarily warm summers, like in 1947, up to five generations of Anopheles developed (Heinz 1950).

As in *case study* 1, the risk modelling presented in *case study* 2 refers to *P. vivax* because it is most relevant in Germany (Dalitz 2005; Jetten and Takken 1994; Maier et al. 2003; Martini 1920; Weyer 1940). *An. atroparvus* occurs mainly in coastal regions in sea-, brackish- and freshwater (Swellengrebel et al. 1935) but is generally found throughout Germany, e.g. in the Black Forest, Leipzig and Thuringia (Mohrig 1969) as well as in the conurbations of Berlin and Frankfurt (Maier et al. 2003).

3.2.2 Calculation of Mean Monthly Air Temperatures

Data on mean monthly air temperatures were provided by the German Meteoro-logical Service (DWD) for the climate reference period 1961–1990 and the period

1991–2007. These local measurements were transformed to surface maps by means of regression kriging (Odeh et al. 1995). This was performed by correlating temperature values and altitudes of the respective sampling sites. The correlation coefficients indicated a pronounced ($r = 0.75$–0.83) and significant relationship. The regression model was used to calculate high-resolution temperature maps based on the global digital elevation model GLOBE[2] (spatial resolution $= 1 \times 1\ km^2$) for each month and period in a geographical information system (GIS). Finally, residual maps on the differences between measured and modelled temperatures were calculated by using ordinary kriging and then subtracted from the regression kriging maps in a GIS to account for over- or rather underestimations. The quality of the surface estimations was assessed by means of cross-validation (Johnston et al. 2001).

Temperature data for the future climate reference periods 1991–2020, 2021–2050 and 2051–2080 were derived from the climate projections REMO[3] (Regional Model, Max Planck Institute for Meteorology) and WettReg[4] (Weather Condition-based Regionalisation Method, Climate and Environment Consulting Potsdam) which are based on the global ECHAM climate model. Both REMO and WettReg were integrated considering two SRES scenarios: Scenario A1B assumes a rapid economic growth, a global population reaching 9 billion in 2050 and then showing a gradual decline, the quick spread of new and efficient technologies and a balanced use of all energy sources; scenario B1 assumes a more integrated, and more ecologically friendly world with rapid economic growth as in A1, but also rapid changes towards a service and information economy, a population increasing up to 9 billion in 2050 and then declining as in scenario A1B, but with reductions in material intensity and the introduction of clean and resource efficient technologies as well as an emphasis on global solutions to economic, social and environmental stability[5] (Houghton et al. 2001). Compared with WettReg, REMO estimates higher temperatures for both historical and future periods. The spatial resolution of the REMO and WettReg maps was $12 \times 12\ km^2$.

3.2.3 *Reproduction of* Plasmodium vivax *in* Anopheles atroparvus

The maps derived from measured historical, recent and future air temperatures were used to calculate the potential temperature-dependent spread of the malaria pathogen *P. vivax* hosted by *An. atroparvus*. For each relevant month, the possible number of secondary infections was estimated using the basic reproduction rate

---

[2] http://www.ngdc.noaa.gov/mgg/topo/globe.html.

[3] http://www.mpimet.mpg.de/home.html.

[4] http://www.cec-potsdam.de.

[5] http://www.ipcc.ch/ipccreports/sres/emission/index.php?idp=30.

(R_0) formula (Lindsay and Thomas 2001; Martens et al. 1999; Smith and McKenzie 2004; Snow et al. 1990) as explained in Sect. 2.2.5. The calculation of R_0 provides the number of secondary infections caused by a single infected Anopheles individual when it meets a potential host population in which every member is susceptible to the agent. In case of $R_0 \geq 1$, malaria is spreading, in case of $R_0 < 1$ there is no such risk (Martens et al. 1999). Adding up the monthly maps on R_0 in a GIS resulted in according maps of possible transmission gates throughout the year for the respective period in the past (1961–1990, 1991–2007) and in future (1991–2020, 2021–2050 and 2051–2080). For calculation, each grid cell of the according monthly R_0 map was coded by a '1' ($R_0 \geq 1$) or a '0' ($R_0 < 1$). At the end, all these monthly grids were summed up in a GIS indicating the number of months where transmission was possible for each grid cell.

3.3 Results

Table 3.1 summarises the results of the modelling of R_0 for each month in the respective climate period regarding both climate models (WettReg, REMO) and both emission scenarios (B1, A1B) for future temperature development. In July and August, most of the territory in Germany was susceptible for malaria transmission in all climate periods investigated. The same goes for June for most periods. The climate reference period 1961–1990 and the WettReg projections for 1991–2020, however, showed only a share of susceptible regions around or below 50 %. For May, only for the REMO scenario A1B during 2051–2080 there were larger areas (34.5 %) showing R_0 values of above 1 implicating that a malaria spread would be possible in that month. In September, malaria spread would be possible to greater extent for both REMO scenarios in 2051–2080 (97.6, 30.5 %) and also in the period 2021–2050 assuming scenario A1B (32.4 %). Finally, only for REMO A1B in 2051–2080, there were some pixels indicating that a spread would be also possible in October (0.04 %).

Figures 3.1, 3.2, 3.3 and 3.4 spatially discriminate for each period and scenario where a malaria spread would be possible in August. The according maps illustrate that due to the higher temperatures predicted by REMO in each period the percentage of susceptible regions was higher than for the respective WettReg scenarios. The same holds true when comparing the modelling results for the two emission scenarios in each future period: Mostly, the A1B scenarios cause larger areas at risk than those based on the B1 scenarios. When comparing the results for the current observation period 1991–2007 with the projections for 1991–2020, it can be stated that in all months the percentages of areas with R_0 above 1 are remarkably higher in the observation period 1991–2007. This may rely on too moderate assumptions made for the IPCC scenarios in this period resulting in too low temperature projections.

By adding up all maps on monthly R_0 values, transmission maps were derived for each period and the respective climate model. Accordingly, Table 3.2 gives a

Table 3.1 Coverage (%) of territory in Germany with a modelled reproduction rate (R_0) of above 1 for the months May–October according to the respective period and climate scenario

Period/scenario	Percentage of coverage of territory in Germany with $R_0 > 1$					
	May	June	July	August	September	October
1961–1990 observed	0	39.7	91.6	85.0	0	0
1991–2007 observed	0.3	71.3	98.1	97.1	0.8	0
1991–2020 REMO A1B	0	65.5	97.5	92.7	2.0	0
1991–2020 REMO B1	2.4	83.2	97.8	95.3	0.6	0
1991–2020 WettReg A1B	0	47.3	96.4	94.8	0	0
1991–2020 WettReg B1	0	55.3	96.5	96.6	0	0
2021–2050 REMO A1B	0.4	83.5	99.3	98.5	32.4	0
2021–2050 REMO B1	0.9	80.7	98.0	95.3	1.6	0
2021–2050 WettReg A1B	0	71.6	98.1	97.5	1.3	0
2021–2050 WettReg B1	0	58.0	97.6	96.6	1.0	0
2051–2080 REMO A1B	34.5	97.4	99.8	99.6	97.6	0.04
2051–2080 REMO B1	3.3	96.2	99.6	99.1	30.5	0
2051–2080 WettReg A1B	0.3	92.9	99.2	99.3	4.6	0
2051–2080 WettReg B1	0	86.0	99.1	98.7	4.2	0

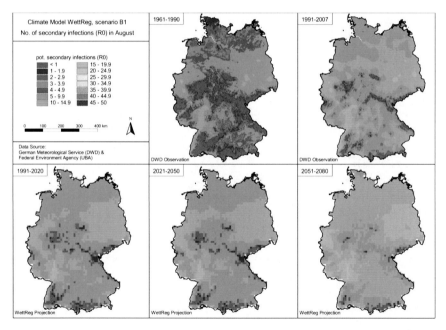

Fig. 3.1 Potential number of secondary infections (R_0) of tertian malaria pathogen in August in Germany according to observed (1961–1990, 1991–2007) and projected (1991–2020, 2021–2050, 2051–2080) air temperatures according to WettReg climate model, emission scenario B1

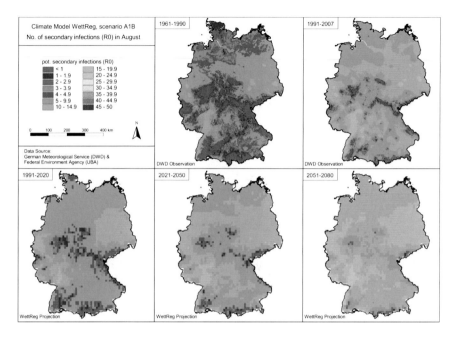

Fig. 3.2 Potential number of secondary infections (R_0) of tertian malaria pathogen in August in Germany according to observed (1961–1990, 1991–2007) and projected (1991–2020, 2021–2050, 2051–2080) air temperatures according to WettReg climate model, emission scenario A1B

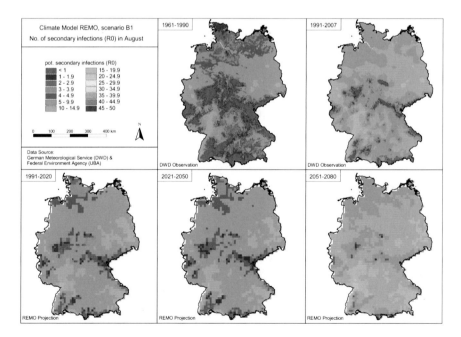

Fig. 3.3 Potential number of secondary infections (R_0) of tertian malaria pathogen in August in Germany according to observed (1961–1990, 1991–2007) and projected (1991–2020, 2021–2050, 2051–2080) air temperatures according to REMO climate model, emission scenario B1

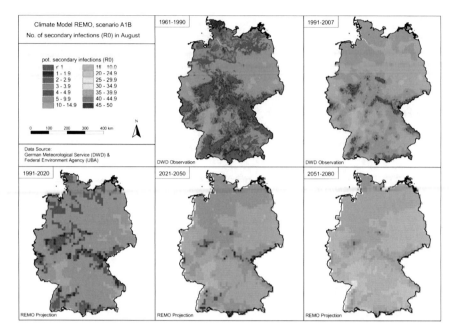

Fig. 3.4 Potential number of secondary infections (R_0) of tertian malaria pathogen in August in Germany according to observed (1961–1990, 1991–2007) and projected (1991–2020, 2021–2050, 2051–2080) air temperatures according to REMO climate model, emission scenario A1B

Table 3.2 Coverage (%) of seasonal malaria transmission gates (months) in Germany (according to Holy et al. 2011)

Period/data	Area coverage (%) of seasonal transmission gates (months)					
	0	1	2	3	4	5
1961–1990 DWD	8.4	6.6	45.3	39.7	0.02	0
1991–2007 DWD	1.8	1.1	25.8	70.3	0.8	0.1
1991–2020 REMO A1B	2.4	5.0	27.1	63.5	2.0	0
1991–2020 REMO B1	1.9	2.5	11.7	81.9	1.7	0.3
1991–2020 WettReg A1B	3.6	1.6	47.5	47.3	0	0
1991–2020 WettReg B1	3.1	0.6	40.9	55.3	0	0
2021–2050 REMO A1B	0.7	0.8	13.8	53.5	30.8	0.4
2021–2050 REMO B1	2.0	2.7	14.6	78.9	1.0	0.7
2021–2050 WettReg A1B	1.9	0.6	25.9	70.3	1.3	0
2021–2050 WettReg B1	2.4	1.0	38.6	57.0	1.0	0
2051–2080 REMO A1B	0.2	0.1	1.1	2.0	62.0	34.5
2051–2080 REMO B1	0.4	0.5	2.1	67.3	26.4	3.3
2051–2080 WettReg A1B	0.7	0.1	6.3	88.3	4.3	0.3
2051–2080 WettReg B1	0.9	0.4	12.7	81.8	4.2	0

detailed overview of the percentages of possible lengths of seasonal transmission gates in Germany for each period and each scenario. For almost all periods, a transmission window of 3 months is most frequent. Only during the climate reference period 1961–1990 a transmission gate of 2 months (45.3 %) dominated, and only in this period a considerable area (8.4 %) was observed where no transmission would be possible, especially in the high and low mountain ranges.

Figures 3.5, 3.6, 3.7, 3.8, 3.9, 3.10, 3.11 and 3.12 illustrate the results condensed in Table 3.2 in the form of transmission maps and according regionalised maps on the respective differences between two following periods. In the climate reference period 1961–1990, the potential seasonal transmission gate of tertian malaria ranged between 0 and 4 months (upper left, Figs. 3.5, 3.7, 3.9, 3.11). At the mountain margins, a transmission would have been possible during 1 month (coverage 6.6 %), whereas the coastal regions, the foothills of the Alps and the Frankish Tableland showed a seasonal transmission gate of 2 months (coverage 45.3 %). With 4 months, the longest transmission period could be found at the Kaiserstuhl and at a local spot in the Upper Rhine valley near Baden–Baden (0.02 %). The remainder of the German territory (39.7 % of the German territory) featured a possible transmission period of 3 months.

In the more recent observation period 1991–2007 (upper right, Figs. 3.5, 3.7, 3.9, 3.11), the maximum seasonal transmission gate already extended to 5 months. In this period, no transmissions were possible only in the upper regions of the low mountain ranges Harz, Rothaargebirge, Thuringian Forest, Erzgebirge, Fichtelgebirge, Bavarian Forest, Black Forest and in the Alps (coverage 1.8 %). The mid-altitudes of the lower mountain ranges featured already possible transmission periods between 1 (coverage 1.1 %) and 2 months (coverage 25.8 %). The coastal regions still allowed a transmission during 2 months, yet the border to the region with a 3 months transmission period (coverage 70.3 %) had moved closer to the coast. Transmissions during 4 months (coverage 0.8 %) would have been possible at single spots in the Leipzig lowlands, the Rhineland and in the Upper Rhine valley. The maximum transmission period allowing for 5 months of possible transmission was also approached at four spots in the Upper Rhine valley (coverage 0.1 %).

Considering the A1B scenarios, both the REMO and the WettReg calculations underestimated the length and the extent of the possible seasonal transmission gate for the climate reference period 1991–2020 compared with the period 1991–2007 that was based on measured values (lower left maps in Figs. 3.8, 3.12). The same goes for the B1 scenario estimated by WettReg (Fig. 3.6). These difference maps depict spatially differentiated that the climate projections predict cooler temperatures than those observed for the last years indicating that the latest temperature rise already surpasses the predicted temperatures. Only the REMO scenario B1 (Figs. 3.9, 3.10) showed a transmission gate prolonged by 0.1 % for the period 1991–2020 in contrast to 1991–2007. Comparing REMO and WettReg estimations, the transmission gates for future periods (2021–2050, and 2051–2080) calculated by REMO showed larger areas at risk than those calculated by WettReg since the WettReg estimations predict a more moderate temperature rise.

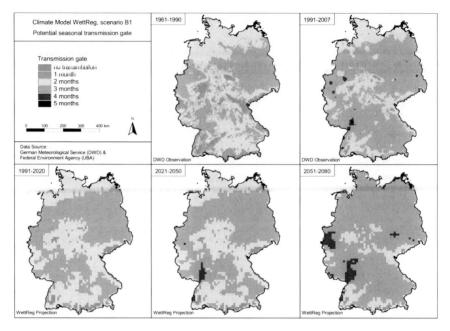

Fig. 3.5 Potential seasonal transmission gates for tertian malaria in Germany according to observed (1961–1990, 1991–2007) and projected (1991–2020, 2021–2050, 2051–2080) air temperatures according to WettReg climate model, emission scenario B1

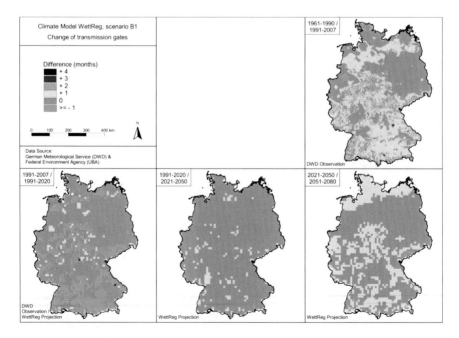

Fig. 3.6 Differences in the length of the potential seasonal transmission gates for tertian malaria in Germany according to observed (1961–1990, 1991–2007) and projected (1991–2020, 2021–2050, 2051–2080) air temperatures according to WettReg climate model, emission scenario B1

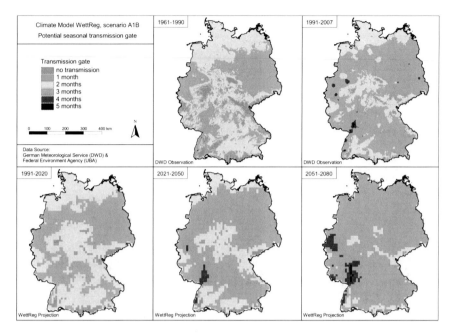

Fig. 3.7 Potential seasonal transmission gates for tertian malaria in Germany according to observed (1961–1990, 1991–2007) and projected (1991–2020, 2021–2050, 2051–2080) air temperatures according to WettReg climate model, emission scenario A1B

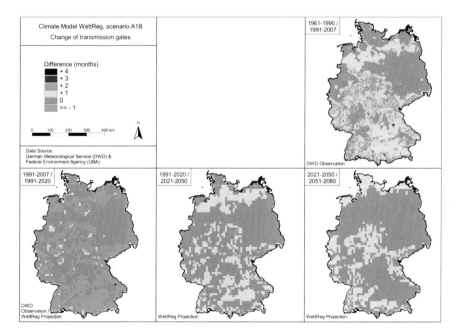

Fig. 3.8 Differences in the length of the potential seasonal transmission gates for tertian malaria in Germany according to observed (1961–1990, 1991–2007) and projected (1991–2020, 2021–2050, 2051–2080) air temperatures according to WettReg climate model, emission scenario A1B

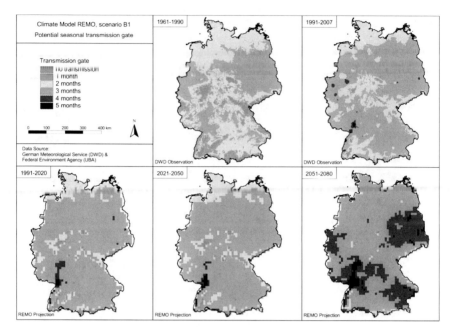

Fig. 3.9 Potential seasonal transmission gates for tertian malaria in Germany according to observed (1961–1990, 1991–2007) and projected (1991–2020, 2021–2050, 2051–2080) air temperatures according to REMO climate model, emission scenario B1

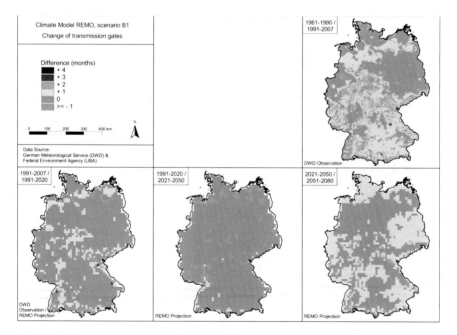

Fig. 3.10 Differences in the length of the potential seasonal transmission gates for tertian malaria in Germany according to observed (1961–1990, 1991–2007) and projected (1991–2020, 2021–2050, 2051–2080) air temperatures according to REMO climate model, emission scenario B1

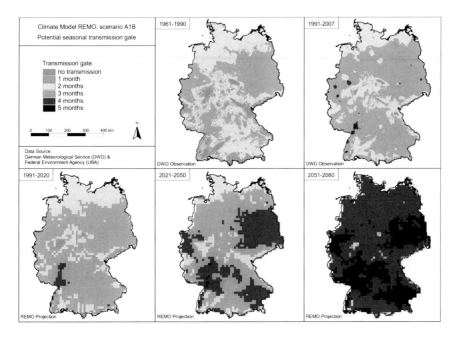

Fig. 3.11 Potential seasonal transmission gates for tertian malaria in Germany according to observed (1961–1990, 1991–2007) and projected (1991–2020, 2021–2050, 2051–2080) air temperatures according to REMO climate model, emission scenario A1B

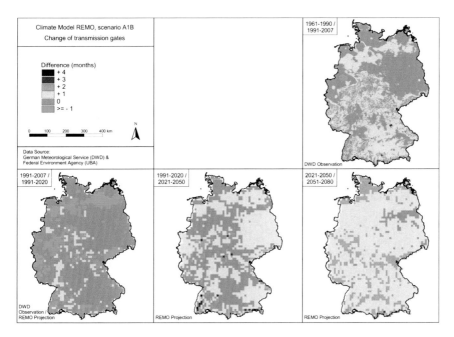

Fig. 3.12 Differences in the length of the potential seasonal transmission gates for tertian malaria in Germany according to observed (1961–1990, 1991–2007) and projected (1991–2020, 2021–2050, 2051–2080) air temperatures according to REMO climate model, emission scenario A1B

During the period 2051–2080, the estimations based on the WettReg scenario B1 (lower right, Fig. 3.5) showed that no transmission would only be possible in the upper parts of the Erzgebirge and the Black Forest as well as in the Alps (0.9 % coverage), the temperatures in the surrounding lower mountain ranges allowed up to 2 months of transmission (0.4, 12.7 % coverage). Four months of possible transmission (4.2 % coverage) were found in the lowlands near Leipzig, in the Rhineland, in the Saarland, in the Upper Rhine valley and in the Rhine-Main-Lowlands. The remainder of Germany was dominated by a 3-months transmission period (81.8 % coverage). Considering the period 2051–2080 as well, WettReg A1B (lower right, Fig. 3.7) showed almost the same regions with no possible transmissions as described for scenario B1. Only those regions located in eastern Germany were dropped. Along the Upper Rhine valley now even 5 months of transmission were possible (0.3 % coverage).

The REMO scenario B1 for the period 2051–2080 (lower right, Fig. 3.9) showed no possible transmissions in the upper Harz Mountains and the Alps (coverage 0.4 %). Areas allowing malaria transmissions during 4 months are mainly found in eastern Germany (Brandenburg and Saxony), the Rhineland and south-eastern Germany and within the Rhine-Main-Lowlands (coverage 3.3 %). The main part of Germany showed a 3-months transmission period (coverage 67.3 %). The estimations for the REMO A1B scenario (lower right, Fig. 3.11) showed comparable spatial patterns, although the transmission periods have been prolonged by 1 month: The percentage of those areas indicating a transmission gate of 4 months was then 62 % instead of 26.4 % for B1, and the areas with a 3-month period decreased from 67.3 to 2 % (Table 3.2). As Fig. 3.11 (lower right) shows, almost the whole territory of Germany features either 4 or 5 (coverage 34.5 %) months of possible malaria transmission.

For a regionalised view, the maps on the modelled transmission gates were intersected by a map on German ecoregions in GIS (Schröder and Schmidt 2001) (Sect. 2.2.4). In Table 3.3, the results of this analysis are exemplarily shown for the WettReg A1B scenario in the period 2051–2080.

At first, one can see that the most frequent length of the transmission window was 3 months. Only in the ecoregions 18, 19, 30, 47 and 63 (see Fig. 2.1) malaria spread was expected to last 4 or 5 months. On the other hand, in the upper low (ecoregion 12) and high mountain ranges (ecoregion 54), there were also regions left where no or just 1 month of transmission was possible. Figures 3.13 (ecoregion 12) and 3.14 (ecoregion 47) illustrate this exemplarily by bar charts depicting the percentages of transmission gates for each period and climate scenario.

3.4 Discussion

Case study 2 modelled the potential spread of tertian malaria by anopheles mosquitoes during 1991–2020, 2021–2050 and 2051–2080. Like for the pilot study for the federal state of Lower Saxony (*case study* 1), in *case study* 2 as well air

Table 3.3 Coverage (%) of seasonal malaria transmission gates (months) for the period 2051–2080 (WettReg A1B) spatially differentiated by German ecoregions

Ecoregion	Total share (%)	0	1	2	3	4	5
8	4.0	0.0	0.0	0.0	78.2	17.8	0.0
12	6.3	0.6	1.9	13.6	70.1	3.2	2.6
18	5.8	0.0	0.0	0.0	7.2	85.6	5.8
19	3.9	0.0	0.0	0.0	22.1	75.6	2.3
20	3.8	0.0	0.0	1.1	86.2	5.7	0.0
22	1.6	0.0	0.0	2.6	46.2	5.1	0.0
26	2.4	0.0	0.0	0.0	92.6	7.4	0.0
30	3.0	0.0	0.0	0.0	46.5	52.3	1.2
42	7.3	0.0	0.0	0.6	84.3	3.9	0.0
43	7.2	0.0	0.0	1.7	95.5	0.0	0.6
46	4.1	0.0	0.0	2.2	85.9	0.0	0.0
47	8.1	0.0	0.0	0.0	49.0	29.9	13.7
54	1.7	4.8	2.4	14.3	69.0	9.5	0.0
55	1.7	0.0	0.0	0.0	93.6	6.4	0.0
56	6.7	0.0	0.0	0.6	61.1	38.3	0.0
57	2.8	0.0	1.4	0.0	82.2	13.7	1.4
58	2.7	0.0	0.0	1.5	88.1	10.4	0.0
62	14.0	0.0	0.0	0.3	71.3	23.5	3.1
63	2.8	0.0	0.0	0.0	13.8	52.3	23.1
118	3.8	0.0	0.0	0.0	82.0	18.0	0.0
119	6.2	0.0	0.0	0.0	54.4	42.2	2.7
Average		**0.1**	**0.2**	**1.6**	**65.5**	**25.1**	**3.1**

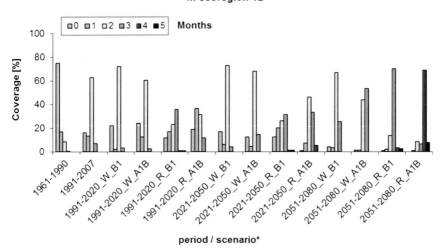

Fig. 3.13 Percentages of the potential seasonal transmission gates for tertian malaria according to observed (1961–1990, 1991–2007) and projected (1991–2020, 2021–2050, 2051–2080) air temperatures spatially differentiated for Germany's ecoregion 12 (upper highlands). *W climate model WettReg, R climate model REMO; *B1/A1B* IPCC emission scenarios

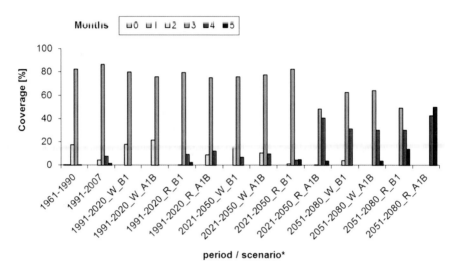

Fig. 3.14 Percentages of the potential seasonal transmission gates for tertian malaria according to observed (1961–1990, 1991–2007) and projected (1991–2020, 2021–2050, 2051–2080) air temperatures spatially differentiated for Germany's ecoregion 47 (Northern Loess Belt and Upper and Lower Rhine Valley). *W climate model WettReg, R climate model REMO; B1/A1B IPCC emission scenarios

temperature data were applied to model the reproduction and development of the mosquito and the malarial parasite. In contrast, future temperature grids for Germany at a scale of 12 km^2 were available from two different climate models: The Regional Model (REMO) by Max Planck Institute for Technology, Hamburg and the Weather Condition-based Regionalisation Method (WettReg) by Climate and Environment Consulting, Potsdam (Sect. 3.2.2). For both models, two different IPCC climate scenarios were considered: A1B and B1.

Using current measurement data for 1991–2007, there was no possibility of malarial spread in only 1.8 % of German territory, mainly in mountainous ranges. On the other hand, 70 % of Germany showed a window of 3 months when malaria spread was possible. Only 0.8 % had a transmission window of 4 months, located in the temperature spoilt regions in the Rhineland and the Upper Rhine Valley. Regarding future development, the projected estimates for 2051–2080 showed a distinct increase in the length of time when malaria could spread and in the areas that could experience this. Taking the more extreme scenario (A1B), about only 0.2–0.7 % of Germany would have no possibility of malarial spread, depending on the climate model used. The REMO model predicted a greater spread with 96.5 % of Germany experiencing a 4–5 month transmission window. In comparison, the more temperate WettReg model predicted for the majority of Germany (88.3 %) a 3-month transmission window.

Skipping other relevant influences besides air temperature, one can conclude that in case of the reintroduction and a permanent, autochthonous establishment of *P. vivax*, a potential of a return of tertian malaria in Germany is theoretically conceivable (Dalitz 2005; Krüger et al. 2001). Yet, so far, the number of imported malaria cases in Germany is assessed to be too low for a re-establishment of autochthonous malaria (Maier et al. 2003). In this context, eastern Germany is assessed to be in a state of an instable equilibrium (Dalitz 2005).

References

Bryan JH, Foley DH, Sutherst RW (1996) Malaria transmission and climate change in Australia. Med J Australia 164:345–347

Dalitz MK (2005) Autochthone Malaria im mitteldeutschen Raum. University of Halle, Dissertation

Hackett LW, Missiroli A (1935) The varieties of Anopheles maculipennis and their relation to the distribution of malaria in Europe. Riv Malariologia XIV(1):1

Hartelt K, Pluta S, Oehme R, Kimmig P (2008) Spread of ticks and tick-borne diseases in Germany due to global warming. Parasitol Res 103(1):109–116

Heinz HJ (1950) Neuere Untersuchungen über die Verbreitung von Anopheles maculipennis in Hamburg. Z angew Entomol 31(2):304–333

Holy M, Schmidt G, Schröder G (2011) Potential malaria outbreak in Germany due to climate warming: risk modelling based on temperature measurements and regional climate models. Environ Sci Pollut Res 18:428–435

Houghton JT, Ding Y, Griggs DJ, Noguer M, van der Linden PJ, Dai X, Maskell K, Johnson CA (eds) (2001) The scientific basis. Contribution of working group i to the third assessment report of the intergovernmental panel on climate change. Cambridge University Press, Cambridge

IPCC (Intergovernmental Panel of Climate Change) (2007) Climate change 2007. Synthesis report, Geneva

Jetten TH, Takken W (1994) Anophelism without malaria: a review of the ecology and distribution of the genus Anopheles in Europe. Wageningen Agricultural University Papers, No. 94(5), Wageningen

Johnston K, Ver Hoef JM, Krivoruchko K, Lucas N (2001) Using ArcGIS geostatistical analyst. ESRI, Redlands

Krüger A, Rech A, Su XZ, Tannich E (2001) Two cases of autochthonous Plasmodium falciparum malaria in Germany with evidence for local transmission by indigenous Anopheles plumbeus. Trop Med Int Health 6(12):983–985

Lindsay SW, Thomas CJ (2001) Global warming and risk of vivax malaria in Great Britain. Global Change Hum Health 2(1):80–84

Maier WA, Grunewald J, Habedank B, Hartelt K, Kampen H, Kimmig P, Naucke T, Oehme R, Vollmer A, Schöler A, Schmitt C (2003) Mögliche Auswirkungen von Klimaveränderung auf die Ausbreitung von primär humanmedizinisch relevanten Krankheitserregern über tierische Vektoren sowie auf die wichtigen Humanparasiten in Deutschland. Climate Change 05/03. Umweltbundesamt, Berlin

Martens P, Kovats RS, Nijhof S, de Vries P, Livermore MTJ, Bradley DJ, Cox J, McMichael AJ (1999) Climate change and future population at risk of malaria. Global Environ Chang 9:89–107

Martini E (1920) Anopheles in Niedersachsen und die Malariagefahr. Hyg Rundsch 22:673–677

Martini E (1946) Lehrbuch der medizinischen Entomologie. Gustav Fischer, Jena

Mohrig W (1969) Die Culiciden Deutschlands. Parasitologische Schriftenreihe 18

Mühlens P (1930) Malaria. Neue Deutsche Klinik. Handwörterbuch der Praktischen Medizin mit besonderer Berücksichtigung der Inneren Medizin, der Kinderheilkunde und ihrer Grenzgebiete VII (31):122–149

Odeh IOA, McBratney AB, Chittleborough DJ (1995) Further results on prediction of soil properties from terrain attributes: heterotopic cokriging and regression-kriging. Geoderma 67(3–4):215–226

Ramsdale C, Snow K (2000) Distribution of the genus Anopheles in Europe. Eur Mosq Bull 7:1–26

Schröder W, Schmidt G (2001) Defining ecoregions as framework for the assessment of ecological monitoring networks in Germany by means of GIS and classification and regression trees (CART). Gate to EHS 1(3):1–9

Smith DL, McKenzie FE (2004) Statics and dynamics of malaria infection in Anopheles mosquitoes. Malar J 3:13

Snow RW, Ikoku A, Omumbo J, Ouma J (1990) The epidemiology, politics and control of malaria epidemics in Kenya: 1900–1998. Roll back malaria, resource network on epidemics. World Health Organisation, Geneva

Swellengrebel NH, de Buck A, Kraan MH, van der Torren G (1935) Occurence in fresh and brackish water on the larvae of Anopheles maculipennis atroparvus and messeae in some coastal provinces of the Netherlands. Quarterly Bulletin of the Health Organisation of the League of Nations V(3):280–294

Tseng WC, Chen CC, Chang CC, Chu YH (2009) Estimating the economic impacts of climate change on infectious diseases: a case study on dengue fever in Taiwan. Climatic Change 92:123–140

Weyer F (1940) Malaria und Malariaübertragung in Ostfriesland. Deut Tropenmedizinische Wochenschr 44(1–2)

WHO (World Health Organization) (2005) World malaria report 2005. World Health Organisation, Geneva

Zhang Y, Peng B, Hiller JE (2008) Climate change and the transmission of vector-borne diseases: a review. Asia Pac J Publ Health 20(1):64–76

Chapter 4
Conclusions and Outlook

Globally, air temperature rose by 0.74 °C between 1906 and 2005 and is projected to rise up to 6.4 °C by 2099, according to the Intergovernmental Panel on Climate Change (IPCC 2007). Climate warming can affect the distribution and the intensity of parasitic diseases that are carried by insects and animals (vector-borne diseases). This is because both the vector and parasites that cause the disease usually flourish in increased temperatures where they benefit from accelerated rates of reproduction and development. Malaria is usually thought to be restricted to the tropics and developing countries, but climate change could bring it closer to Europe, especially in countries where it used to be present. Malaria was endemic in Europe until the middle of the last century, when eradication efforts could finally wipe out the disease. But only the pathogen seemed to be expelled, the vector in form of *Anopheles* mosquitoes is still present. Tertian malaria, a sort of malaria, was prevalent in north-western parts of Germany until the 1950s before it was eradicated. But the vector itself (the mosquito) is still present and capable to transmit malaria. Infected people from malarial regions could introduce a new onset of malaria.

Two case studies were conducted to investigate the potential spread of tertian malaria: *Case study 1* dealt with the potential temperature-dependent transmission of malaria in Lower Saxony where recent anopheline larvae findings in Lower Saxony could be matched with past endemic malaria zones. It could be demonstrated that during the periods 1947–1960, 1961–1990 and 1985–2004 in most parts of Lower Saxony the temperature-dependent seasonal transmission gate was 2 months. Projections on the malaria spread in Lower Saxony were performed by using predictions on air temperature rise of the IPCC (2001) for best- and worst-case scenarios covering the years 2020, 2060 and 2100. By rise of air temperature in future, potential transmission gates for tertian malaria could be prolonged from 2 months in average to 6 months in worst case, especially in temperate coastal regions of Lower Saxony. It is just this area where the vectors in form of *Anopheles atroparvus* and *Anopheles messeae* are still present (Wilke et al. 2006) and where malaria tertiana was endemic till the 1950s (Weyer 1956). Since whole Germany was considered to be suitable for the establishment of populations of *Anopheles atroparvus*, *case study* 2 was conducted subsequently to assess areas at risk for malaria spread. In contrast to *case study 1*, for *case study 2* projected maps

W. Schröder and G. Schmidt, *Modelling Potential Malaria Spread in Germany by Use of Climate Change Projections*, SpringerBriefs in Environmental Science, DOI: 10.1007/978-3-319-03823-0_4, © The Author(s) 2014

on air temperatures with a spatial resolution of 12 km^2 were available from IPCC
(2007) for the periods 1991–2020, 2021–2050 and 2051–2080 based on two climate models (REMO, WettReg) and two emission scenarios (A1B, B1), respectively. According to measurement data on air temperature for 1990–1997, 70 % of
Germany, mainly in coastal regions and foothills, had a transmission window of
3 months when malarial spread was possible. However, the projected estimates for
2051–2080 showed an increase in the length of the seasonal transmission gates.
Using temperature data derived by the REMO model, 96.5 % of Germany would
experience a 4–5 month transmission window assuming the more extreme emission scenario A1B. According to the more conservative WettReg model, the
majority of Germany (88.3 %) is expected to maintain a 3 month transmission
window for the period 2051–2080 for the same emission scenario.

In conclusion, both studies give evidence that the import of the malaria
pathogens might cause risk of an autochthon malaria transmission, presuming the
Anophelinae occur in high frequency and the breeding places are close to
anthropogenic settlements and other requirements for transmission are met. On the
other hand, prospected air temperature rise should prolong the potential transmission gates of Plasmodium vivax in future. If the pathogen will be introduced
and will still be able to be transmitted by these mosquito species, transmission of
tertian malaria, thus, can take place in Germany.

The introduction of pathogens to Germany is not a hypothetical issue: There
were 150 cases of imported malaria reported from 1999–2003 amounting to
24.3 % of all cases registered throughout Europe (Mühlberger et al. 2004). As a
consequence, several local malaria transmissions were registered in Germany
(Krüger et al. 2001; Zoller et al. 2009) and other European countries such as Italy
(Gubler et al. 2001). Further local transmissions of Plasmodium vivax were
recorded in the United States which were—in fact—considered being malaria free
(Malecki et al. 2003; Pastor et al. 2002). One possible way of reintroduction of the
malaria pathogen could be via infected travellers arriving from current endemic
malaria regions. The high number of airport and baggage malaria cases in
Germany proves that this possibility is not only hypothetical (Gratz et al. 2000;
RKI 1999). In these densely populated areas a higher proportion of immigrants
and/or tourists may import malaria pathogens from their home countries/holiday
destinations (Doudier et al. 2007; Millet et al. 2008). In this context, the Frankfurt
airport as an international hub is of particular importance because of the high
numbers of passengers arriving from endemic malaria regions. Between 2001 and
2006, 4,639 imported cases of malaria were registered in Germany. 78 % of the
628 infections registered in 2005 originated from the malaria tropica pathogen
Plasmodium falciparum, 12 % were Plasmodium vivax infections, 4 % Plasmodium ovale and 3 % Plasmodium malariae. Six infected people died in 2005 (Ebert
and Fleischer 2008). Two autochthonous malaria cases are reported from a
German hospital which could be traced back to infections by the indigenous
mosquito species Anopheles plumbeus which had taken blood meals from an in-
patient from Angola with a chronic Plasmodium falciparum infection. As the
infections occurred in August 1997 showing mean daily temperatures between 21

and 27 °C, a transmission by the vector *Anopheles plumbeus* was assumed (Krüger et al. 2001). In case of the so-called airport and baggage malaria, the infections by mosquitoes coming as stowaways happen not only in airplanes or airport buildings but also kilometres afar from airports so that people who have never travelled to endemic malaria regions are at risk, too (RKI 1999).

The prediction of environmental change and the resulting health impacts encounter several difficulties. Some of them are of scientific nature and refer to a deficient understanding of the actual processes. Some of the uncertainties are related to the modelling of the relation between the increase in temperature and humidity on the one hand, and mosquito breeding, survival and biting behaviour on the other. Other uncertainties pertain to what can be foreseen about future emissions of greenhouse gases. And, finally, there is further uncertainty caused by data collection. Accordingly, the effects of climate change on the spread of vector-borne diseases are subjects of numerous studies and are discussed controversially (Hoshen and Morse 2004; Omumbo et al. 2004; Martens et al. 1999). Concerning the coastal zone of Lower Saxony as a former malaria region, an interdisciplinary project at the University of Bremen is dealing with climate change and its consequences. A regional climate scenario was calculated, and a temperature increase of 2.7 °C for 2050 could be determined.[1] Also for the last 20 years, higher temperatures at ground level could be found which substantiate the results of the investigation at hand. But there is no general correspondence that rising temperatures lead to a progression of the disease towards northern latitudes. Reiter (2000) and Small et al. (2003) concluded that temperature is not the predominant factor influencing malaria transmission and recommended that non-climatic factors should be examined closely. Other studies yield that the rise in temperature will affect the spread of malaria in Europe (Lindsay and Thomas 2001). Accordingly, Schmidt et al. (2008) as well as Schröder and Schmidt (2008) identified Lower Saxony (Northern Germany) as an area at risk for secondary malaria infections in terms of thermal conditions. Moreover, the risk of an introduction of more potent malaria vectors from the Mediterranean region in Germany is named as a consequence of rising temperatures (Maier et al. 2003: 189–190). However, as there is no explicit evidence so far that climate warming will not influence the distribution of malaria, there is a need for research in potential areas at risk (e.g. Lower Saxony). For a more accurate spatiotemporal analysis of the risk potential, further influencing factors should be considered, e.g. by including maps on the distribution of natural and artificial water bodies (wetlands and river networks), precipitation, humidity, as well as on population exposure and livestock density. Therefore, further investigations referring to breeding site preferences should be conducted. To this end, Wilke et al. (2006) tried to find correlations between certain habitat characteristics, like pH, chlorine content or vegetation structure and the distribution of different Anophelinae species. The conclusions are of major interest for a more detailed analysis concerning the areas at risk. By knowing certain

---

[1] http://www.klimu.uni-bremen.de/english/klimaszenarioenglisch.html.

preferences of a species-dependent selection of breeding places, the areas at risk detected in this investigation could be examined in terms of those preferences. Those of course are different from place to place and according to this, Lower Saxony can be divided into several bio-geographical regions with different climatic conditions, soil types and types of aquatic habitats. An identification pattern concerning the respective areas has to be created first. Here the BOG-Archive can contribute much additional information about Lower Saxony's waters.

The model applied in *case study 1* (Lower Saxony) and *case study 2* (Germany) was based on the basic reproduction rate (R_0) which is well established in malaria transmission risk assessment (Dietz 1993; Lindsay and Thomas 2001; Martens et al. 1999; Snow et al. 1999). The formula takes the daily survival probability of the examined species into its calculation which is determined particularly by temperature conditions (MacDonald 1956). However, the formula has also some deficits. The relationship between climate and transmission potential of malaria is still only partly understood. Furthermore, in addition to meteorological factors, disease transmission dynamics are influenced by many other drivers with regard to the hosts, to the vectors and to other members of the affected ecosystems and interactions between them (Martens et al. 1999). Certain variables which can play a significant role in the complex dynamic process of disease spread were excluded, such as precipitation, humidity, availability of mosquito breeding sites and other ecological factors determining the developmental process of the vector mosquitoes and also hygiene and medical statuses were faded out (Martens et al. 1999; Leemans 2005). If temperature was the only relevant parameter, malaria would not have become eradicated in Europe, as demonstrated by the risk modelling. The impact of climate change poses enormous challenges to scientists because of the considerable amount of complexity and uncertainty: The same environmental change can have quite different effects in different places or times. So there is a need for investigations on the short- and long-term dynamics of complex systems, and this requires an interdisciplinary approach integrating, amongst others, the fields of biogeography, ecology and both medical and social sciences. Considering future implications, it has also stated that the respective trends of the REMO and WettReg projections which are based on different emission scenarios show a wide range of temperature development.

Several models have been developed to describe a specific bioclimatic envelope for malaria. Those of them that match the presence of a particular species with a discrete range of temperature and precipitation parameters can be used to project the effect of climate change on the spatial patterns of vector distribution (Lindsay et al. 1998). As demonstrated in this investigation and those of Hay et al. (2002) and Patz et al. (2002), grid maps interpolated from meteorological data often provide the basis for modelling the inter-relationship among vectors, vector-borne diseases and climate. However, such approaches cannot include all factors that affect the distribution of species. For example, local barriers and interactions between species are important factors that determine whether species colonise suitable habitats (Davis et al. 1998). Martin and Lefebvre (1995) could estimate the changes in malaria risk based on moisture and minimum and maximum temperatures required for parasite development. The modelling results fit well with the distribution of malaria in the nineteenth century and in the 1990s. Rogers and Randolph (2000) found that malaria will increase in some areas and decrease in others by 2050. Martens et al. (1999) developed a model that is basically similar to the approach applied for modelling the temperature-dependent increase of the potential malaria transmission in Lower Saxony and Germany: Climate-change scenarios were linked with the formula for the basic reproduction rate (R_0) to calculate the epidemic potential of a malaria-vectoring mosquito population. The basic reproduction rate is defined as the number of new cases of a given disease that will arise from one case when introduced into a non-immune host population during a single transmission cycle (Anderson and May 1991). Model variables that are sensitive to temperature include mosquito density, feeding frequency, survival and extrinsic incubation period. The extrinsic incubation period, i.e. the development of the parasite in the mosquito, is of particular importance. The minimum temperature for parasite development is the limiting factor for malaria transmission in many areas.

Malaria incidence is strongly influenced by geography and prevailing climate. However, malaria incidences correlate with per capita income, too. Apportioning malaria causality between environment and socio-economic conditions is, therefore, problematic (Martens and Thomas 2005). The problem of modelling the impacts of global change processes on human health is that it has to cope with a variety of processes that operate on different temporal and spatial levels and differ in complexity: Modelling has to link processes that differ by nature, such as physical processes, monetary processes, social processes and policy processes. Furthermore, modelling has to deal with the fundamental problem of predicting across temporal and spatial scales. Despite such limitations of modelling, models draw attention to the climate change-related impacts and point out the relative importance of the factors. This could help to increase the understanding of the impacts of climate change, and to identify gaps in data and knowledge needed to improve the analysis of these effects.

Global change is likely to affect human and animal health via complex interactions that may affect a vector-borne disease: the mosquito vector, the parasite, human hosts, climate, topography, vegetation, aquatic habitats, demography, health care and socio-economic status. Whether the pilot study would have been focused not on malaria but on dengue or leishmaniasis the conclusions and recommendations would not have been much different. Environmental change such as climate change will affect the incidence of vector-borne diseases remarkably. The uncertainty of these impacts results from the level of change as well as from the spatial and temporal patterns. Until recently, it seemed difficult to provide specific evidence that environmental change had caused a significant positive impact on a vector-borne disease, leading to increased cases of disease. Takken et al. (2005) provided evidence that environmental degradation, accelerated by increased precipitation and higher temperatures, favour vector-borne diseases. Hence, it is expected that transmission will increase in those countries which are too poor to provide adequate health care. By contrast, malaria is unlikely to return to former endemic countries as for example Germany, because the high standard of living and because environmental measures prevent a return of malaria endemicity, even under climate change (Rogers and Randolph 2000). It is generally accepted that transmission of many infectious diseases is affected by climatic conditions. Diseases caused by pathogens which spend part of their life cycle outside of human or other warm-blooded hosts are particularly climate sensitive. Some of these diseases are among the most important global causes of mortality and morbidity, particularly in low-income societies in developing countries. In many environments, these diseases occur as epidemics, possibly triggered by changes in climatic conditions favouring higher transmission rates. Following Takken et al. (2005), global environmental changes will cause both increases and decreases in the areas appropriate for vector-borne diseases transmission, but a bunch of factors will determine the susceptibility of respective countries for these diseases. Assessments on the spatial and temporal patterns should be elaborated from both an environmental and a socio-economic point of view.

Together with several studies (Takken et al. 2005), this investigation could prove geostatistics, ecoregionalisation and GIS to be powerful tools for mapping the spatial patterns of potential malaria transmissions and for assessing the associated ecological factors in terms of an ecological land classification. A better understanding of the malaria transmission helps to design or to optimise both monitoring and control programmes. GIS has already been applied for malaria surveillance as well as for malaria prevention (Gimnig et al. 2005; Kitron et al. 1994). Referring to the compilation given by Takken et al. (2005), the approach presented might be classified as sophisticated and competitive. Koslowsky (2002) evaluated the risk of importing Bluetongue Disease by *Culicoides* into Germany with the help of a GIS. But this study was only based on climatic aspects but neither considered a mathematical model nor the characteristics of the respective species into the risk assessment. Epidemiological research including landscape ecological approaches contributes to an appropriate risk assessment and can help to initiate preventative measures. Areas at risk can be determined by means of GIS and remote sensing techniques and then be included in action plans (Gimnig et al. 2005; Ponçon et al. 2008; Tran et al. 2008). Beyond this, the focus should not only be concentrated on humans but on livestock as well, especially for those regions where cattle or poultry is kept in high densities, like Lower Saxony in northern Germany. As demonstrated by Schmidt et al. (2008), a differentiation of environmental preconditions affecting the malaria spread should be performed by intersecting the ecoregions of Germany (Schröder and Schmidt 2001) with transmission maps. The same approach is applicable also for European-wide investigations of malaria spread by using a map depicting the ecoregions of Europe (Hornsmann et al. 2008; Pesch et al. 2011).

As climate change will probably confront us with increased numbers of vector species, a systematic and consistent vector monitoring should be initiated to improve the spatial differentiation and, thus, the quality of the risk assessment.

Socio-scientific investigations on ageing structure, leisure and recreation activities as well as on livestock husbandry could further specify the information on potential host populations in space and time. Sensitivity analyses could support the assessment of the relevance of different known factors for the calculation of the reproduction rate, like the number of blood meals or the lifespan of the vector.

WHO (2004) evaluated the potential of climate-based disease early warning systems (EWS) as an instrument of improving preparedness for, and response to, epidemics. Climate-sensitive diseases were identified as being of major public health importance. Thus, WHO (2004) evaluated the current state of the art in climate-based modelling of these diseases, as well as future requirements and recommendations. Climate-based disease EWS go back to Gill (1921, 1923). Nowadays, in times of climate change and since climate and other environmental data, GIS and other tools required to link these observations with disease data have become widely available, there is clear justification for investigating the potential of climate-based EWS regarding their usefulness in planning of control interventions. Following WHO (2004), no large-scale EWS is yet in place but for some diseases, such as malaria and Rift Valley fever, early warning systems based on climatic conditions are going to be used in selected locations to alert ministries of health to the potential for increased risk of outbreaks and to improve epidemic preparedness. However, the use of such models is just beginning, and experience with their use is limited for several reasons: Affordable and accessible data as well as analytical tools have become widely available just recently. Therefore, the development of EWS is at a relatively early stage. As few studies have been published there are no generally agreed criteria for accessing predictive accuracy. Most EWS have not been tested in locations outside of the original study area, focus solely on climate factors and do not encompass other influences. Therefore, it is difficult to evaluate quality and utility of existing systems. High-quality, long-term disease and environmental data must be available for the development of models relating climate and other environmental factors to infectious disease. Up to now, disease and climate modelling has been restricted to discrete datasets for relatively small areas. This is useful for methodological reasons but there are relevant questions left concerning the extent to which findings from these studies can be generalised spatially and the assessment of the prediction accuracy. Predictive models should be tested in locations covering a broad range of ecological conditions. Non-climatic influences on variations in disease rates should be investigated, too.

EWS should be developed according to the principles of public health surveillance (Brookmeyer and Stroup 2004; Teutsch and Churchill 1994; Waller and Gotway 2004) as an Internet-based geographical information system (WebGIS) using open source software (Peng and Tsou 2003). The Open Source Initiative (OSI) specifies the concept of open source by several criteria (Williams 2002). The access to the source code has to be free without any constraints in terms of circulation of the software to third parties or certain users and the range of use. The licence does not allow the discrimination of anyone. Mostly, the software is available and downloadable via the Internet and modifications of the source code have to be transferred under the same terms of use. It is allowed to use parts of the source code in other free software products. It is only allowed to use the concept open source if the software is protected by one of the licence models of the OSI. The most commonly used licence of the Free Software Foundation is the GNU Public Licence. It is not allowed to demand money for the acquisition and the use of the software. This does not include a fee for the installation or user-based modifications of the software. Nowadays, open source is an alternative compared to proprietary software but without any warranty for the function of the software. Users are not only people at home offices, but also companies and public authorities that are using open source software in all fields of information technology. Examples are operating systems for servers like Linux and server-based software like http-servers or content management systems and desktop GIS software like GRASS GIS[2] or JUMP.[3] Proprietary software products have their own licences and copyrights (Spath and Günther 2005). For assembling the WebGIS, the standards of the Open Geospatial Consortium (OGC) have to be considered as well as the access to spatial and measured data and metadata of the disease surveillance and environmental monitoring programmes. The standards of the OGC constitute the base of interoperable networks for spatial data infrastructures and additional attribute data located anywhere in the world. They are designed by using ISO-, CEN- and other standards as basis and implemented into the most GIS software (Korduan and Zehner 2008). The most important standards of the OGC are the open web services (OWS) which are interfaces for a standardised access to remotely located geodata (Müller and Augstein 2005). An example for OWS is the web feature service (WFS) which provides vector-based geodata instead of raster data like a web map service (WMS) does. The catalogue service offers metadata management (Müller and Augstein 2005). The simple features implementation specifications (e.g. SQL) define interfaces which provide a transparent access to geodata in heterogeneous and data publishing systems. Another important standard for WebGIS is the web processing service (WPS) which allows the standardised use of online GIS functions, for example with the open source software GRASS GIS.

It would be wrong to confuse the modelling results with reality. The modelling results should rather be considered as an approximation and a useful tool to determine areas at risk and a pre-condition to control them (Smith and McKenzie 2004). The impact of climate change poses enormous challenges to scientists because of the considerable amount of complexity and uncertainty: The same environmental change may induce quite different effects in different places or times. Thus, there is a strong need for investigations on the short- and long-term dynamics of complex systems, and this requires an interdisciplinary approach integrating, amongst others, the fields of biogeography, ecology, and both medical and social sciences. Some European countries like the U.K. and Italy have already

---

[2] http://grass.osgeo.org.

[3] http://www.vividsolutions.com/jump.

undertaken studies in that field, concluding that climate warming would increase the malaria transmission potential (Lindsay and Thomas 2001). In Germany, so far, there is no systematic risk assessment which in close contact with environmental and epidemiological monitoring aims at identifying hot spot areas where vector-borne diseases are likely to happen. A better understanding of the malaria transmission in a given area helps to design or to optimise both monitoring and control programmes.

References

Anderson RM, May RM (1991) Infectious diseases of humans: dynamics and control. Oxford University Press, Oxford

Brookmeyer R, Stroup D (2004) Monitoring the health of populations: statistical principles and methods for public health surveillance. Oxford University Press, New York

Davis AJ, Jenkinson LS, Lawton JH, Shorrocks B, Wood S (1998) Making mistakes when predicting shifts in species range in response to global warming. Nature 391:783–786

Dietz K (1993) The estimation of the basic reproduction number for infectious diseases. Stat Methods Med Res 2:23–41

Doudier B, Bogreau H, DeVries A, Ponçon N, Stauffer WM, Fontenille D, Rogier C, Parola P (2007) Possible autochtonous malaria from Marseille to Minneapolis. Emerg Infect Dis 13(8):1236–1238

Ebert B, Fleischer B (2008) Malaria: Stellungnahmen des Arbeitskreises Blut des Bundesministeriums für Gesundheit. Bundesgesundheitsblatt—Gesundheitsforschung—Gesundheitsschutz 51:236–249

Gill CA (1921) The role of meteorology on malaria. Indian J Med Res 8:633–693

Gill CA (1923) The prediction of malaria epidemics. Indian J Med Res 10:1136–1143

Gimnig JE, Hightower AW, Hawley WA (2005) Application of geographic information systems to the study of the ecology of mosquitoes and mosquito-borne diseases. In: Takken W, Martens P, Bogers RJ (eds) Environmental change and malaria risk: global and local implications. Springer, Dordrecht

Gratz NG, Steffen R, Cocksedge W (2000) Why aircraft disinsection? Bull World Health Organ 78(8):995–1004

Gubler DJ, Reiter P, Ebi KL, Yap W, Nasci R, Patz JA (2001) Climate variability and change in the United States: potential impacts on vector- and rodent-borne diseases. Environ Health Persp 109(2):223–233

Hay SI, Cox J, Rogers DJ, Randolph SE, Stern DI, Shanks GD, Myers MF, Snow RW (2002) Climate change: regional warming and malaria resurgence-reply. Nature 420:628

Hornsmann I, Pesch R, Schmidt G, Schröder W (2008) Calculation of an ecological land classification of Europe (ELCE) and its application for optimising environmental monitoring networks. In: Car A, Griesebner G, Strobl J (eds) Geospatial crossroads @ GI_Forum '08: proceedings of the geoinformatics forum Salzburg. Wichmann, Heidelberg, 140–151

Hoshen MB, Morse AP (2004) A weather-driven model of malaria transmission. Malar J 3:32

IPCC (Intergovernmental Panel of Climate Change) (2001) Climate change: the scientific basis. Cambridge University Press, Cambridge

IPCC (Intergovernmental Panel of Climate Change) (2007) Climate change 2007. Synthesis report, Geneva

Kitron U, Pener H, Costin C, Orshan L, Greenberg Z, Shalom U (1994) Geographic information system in malaria surveillance: mosquito breeding and imported cases in Israel, 1992. Am J Trop Med Hyg 50:550–556

Koslowsky S (2002) Bluetounge disease in Deutschland? Risikoabschätzung mit Hilfe eines Geographischen Informationssystems (GIS). Dissertation, Freie Universität Berlin

Korduan P, Zehner ML (2008) Geoinformation im Internet. Technologien zur Nutzung raumbezogener Informationen im WWW. Wichmann, Heidelberg

Kruger A, Rech A, Su XZ, Tannich E (2001) Two cases of autochthonous *Plasmodium falciparum* malaria in Germany with evidence for local transmission by indigenous *Anopheles plumbeus*. Trop Med Int Health 6:983–985

Leemans R (2005) Global environmental change and health. Integrating knowledge form natural, socioeconomic and medical sciences. In: Takken W, Martens P, Bogers RJ (eds) Environmental change and malaria risk. Global and local implications. Springer, Dordrecht

Lindsay SW, Thomas CJ (2001) Global warming and risk of vivax malaria in Great Britain. Glob Change Hum Health 2(1):80–84

Lindsay SW, Parson L, Thomas CJ (1998) Mapping the ranges and relative abundance of the two principal African malaria vectors, *Anopheles gambiae* sensu stricto and *An. arabiensis*, using climate data. Proc Roy Soc Lond B, Biol Sci 265:847–854

MacDonald G (1956) Epidemiological basis of malaria control. Bull World Health Org 15(3–5):613–626

Maier WA, Grunewald J, Habedank B, Hartelt K, Kampen H, Kimmig P, Naucke T, Oehme R, Vollmer A, Schöler A, Schmitt C (2003) Mögliche Auswirkungen von Klimaveränderung auf die Ausbreitung von primär humanmedizinisch relevanten Krankheitserregern über tierische Vektoren sowie auf die wichtigen Humanparasiten in Deutschland. Climate Change 05/03. Umweltbundesamt, Berlin

Malecki JM, Kumar S, Johnson BF, Gidley ML, O'Connor TE, Petenbrink J, Bush L, Morand J, Perez MT, Pillai S, Crockett L, Blackmore C, Bradford E, Wirtz RA, Barnwell JW, DaSilva AJ, Causer LM, Parise ME (2003) Local transmission of *Plasmodium vivax* malaria—Palm Beach county, Florida. MMWR 52(38):908–911

Martens P, Thomas C (2005) Climate change and malaria risk: complexity and scaling. In: Takken W, Martens P, Bogers RJ (eds) Environmental change and malaria risk. Global and local implications. Springer, Dordrecht

Martens P, Kovats RS, Nijhof S, de Vries P, Livermore MTJ, Bradley DJ, Cox J, McMichael AJ (1999) Climate change and future population at risk of malaria. Glob Environ Change 9:89–107

Martin PH, Lefebvre MG (1995) Malaria and climate: sensitivity of malaria potential transmission to climate. Ambio 24:200–207

Millet JP, Gercia de Olalla P, Carillo-Santisteve P, Gascón J, Treviño B, Muñoz J, Gomez i Prat J, Cabezos J, Gonzáles Cordón A, Caylà JA (2008) Imported malaria in a cosmopolitan European city: a mirror image of the world epidemiological situation. Malar J 7:56

Mühlberger N, Jelinek T, Gascon J, Probst M, Zoller T, Schunk M, Beran J, Gjørup I, Behrens RH, Clerinx J, Björkman A, McWhinney P, Matteelli A, Lopez-Velez R, Bisoffi Z, Hellgren U, Puente S, Schmid ML, Myrvang B, Holthoff-Stich ML, Laferl H, Hatz C, Kollaritsch H, Kapaun A, Knobloch J, Iversen J, Kotlowski A, Malvy DJM, Kern P, Fry G, Siikamaki H, Schulze MH, Soula G, Paul M, Gómez i Prat J, Lehmann V, Bouchaud O, da Cunha S, Atouguia J, Boecken G (2004) Epidemiology and clinical features of *vivax* malaria imported to Europe: sentinel surveillance data from TropNetEurop. Malar J 3:5

Müller M, Augstein B (2005) Das Hamburger Umweltinformationssystem HUIS—Integration von Umweltdaten auf Basis eines GDI-Ansatzes. In: Fischer-Stabel P (ed) Umweltinformationssysteme. Wichmann, Heidelberg

Omumbo JA, Hay SI, Guerra CA, Snow RW (2004) The relationship between the *Plasmodium falciparum* parasite ratio in childhood and climate estimates of malaria transmission in Kenya. Malar J 3:17

Pastor A, Pastor A, Neely J, Goodfriend D, Marr J, Jenkins S, Woolard D, Pettit D, Gaines D, Sockwell D, Garvey C, Jordan C, Lacey C, DuVernoy T, Roberts D, Robert L, Santos P,

Wirtz R, MacArthur J, O'Brien M, Causer L (2002) Local transmission of *Plasmodium vivax* malaria—Virginia. MMWR 51(41):921–923

Patz JA, Hulme M, Rosenzweig C, Mitchell TD, Goldberg RA, Githeko AK, Lele S, McMichael AJ, Le Sueur D (2002) Climate change: regional warming and malaria resurgence. Nature 420:627–628

Peng ZR, Tsou MH (2003) Internet GIS: distributed geographic information services for the internet and wireless networks. Wiley, Hoboken

Pesch R, Schmidt G, Schröder W, Weustermann I (2011) Application of Cart in ecological landscape mapping: two case studies. Ecol Ind 11:115–122

Ponçon N, Tran A, Toty C, Luty AJF, Fontenille D (2008) A quantitative risk assessment approach for mosquito-borne diseases: malaria re-emergence in southern France. Malar J 7:147

Reiter P (2000) Malaria and global warming in perspective? Emerg Infect Dis 6:438–439

RKI (Robert-Koch-Institut) (1999) Zur Airport-Malaria und Baggage-Malaria. Epidemiologisches Bull 37(99):274

Rogers DJ, Randolph SE (2000) The global spread of malaria in a future, warmer world. Science 289:1763–1766

Schmidt G, Holy M, Schröder W (2008) Vector-associated diseases in the contect of climate change: Analysis and evaluation of the differences in the potential spread of tertian malaria in the ecoregions of Lower Saxony. Ital J Public Health 5(4):245–252

Schröder W, Schmidt G (2001) Defining ecoregions as framework for the assessment of 350 ecological monitoring networks in Germany by means of GIS and classification and 351 regression trees (CART). Gate to EHS 1(3):1–9

Schröder W, Schmidt G (2008) Mapping the potential temperature-dependent tertian malaria transmission within the ecoregions of Lower Saxony (Germany). Int J Med Microbiol 298(S1):38–49

Small J, Goetz SJ, Hay SI (2003) Climatic suitability for malaria transmission in Africa 1911–1995. Proc Natl Acad Sci U S A 100(26):15341–15345

Smith DL, McKenzie FE (2004) Statics and dynamics of malaria infection in *Anopheles* mosquitoes. Malar J 3:13

Snow RW, Ikoku A, Omumbo J, Ouma J (1999) The epidemiology, politics and control of malaria epidemics in Kenya: 1900–1998. Roll back malaria, resource network on epidemics. World Health Organisation, Nairobi

Spath D, Günther J (2005) Open Source Software—Strukturwandel oder Strohfeuer?—Eine empirische Studie zu Trends und Entwicklungen zum Einsatz von Open Source Software in der öffentlichen Verwaltung und IT-Unternehmen in Deutschland. Frauenhofer IAO. http://www.iao.fraunhofer.de/d/oss_studie.pdf. Accessed 17 June 2013

Takken W, Martens P, Bogers RJ (eds) (2005) Environmental change and malaria risk: global and local implications. Springer, Dordrecht

Teutsch SM, Churchill RE (1994) Principles and practice of public health surveillance. Oxford University Press, New York

Tran A, Ponçon N, Toty C, Linard C, Guis H, Ferré JB, Lo Seen D, Roger F, de la Rocque S, Fontenille D, Baldet T (2008) Using remote sensing to map larval and adult populations of *Anopheles* hyrcanus (Diptera: Culicidae) potential malaria vector in Southern France. Int J Health Geogr 7:9

Waller LA, Gotway CA (2004) Applied spatial statistics for public health data. Wiley, New York

Weyer F (1956) Bemerkungen zum Erlöschen der ostfriesischen Malaria und zur *Anopheles*-Lage in Deutschland. Z Tropenmed Parasitol 7:219–228

WHO (World Health Organistion) (2004) Using climate to predict infectious disease outbreaks: a review. World Health Organistion, Geneva

Wilke A, Kiel E, Schröder W, Kampen H (2006) *Anophelinae* (Diptera: Culicidae) in ausgewählten Marschgebieten Niedersachsens: Bestandserfassung, Habitatbindung und Interpolation. Mitt Dtsch Ges Allg Angew Entomol 15:357–362

Williams S (2002) Free as in freedom. Richard Stallman's crusade for free software. O'Reilly, Sebastopol, Cambridge
Zoller T, Naucke TJ, May J, Hoffmeister B, Flick H, Williams CJ, Frank C, Bergmann F, Suttorp N, Mockenhaupt P (2009) Malaria transmission in non-endemic areas: case report, review of the literature and implications for public health management. Malar J 8:71

Printed by Publishers' Graphics LLC
DBT140218.15.17.138